Pastoralism, Uncertainty

Praise for this book

A critically important and timely book. It explains why pastoralists are experts in managing uncertainty, and why far more participatory, context-specific analysis is needed if the dismal performance of decades of 'pastoral development' is to be reversed.

Andy Catley, Tufts University, Boston

This book is a timely and much needed window into the resilience of pastoralists worldwide, offering important insights into how to increase adaptive capacity in the face of climate change. It offers further proof that pastoralism is not an historical anachronism, but a sustainable solution for both people and the Earth.

Maryam Niamir-Fuller, co-chair of International Support Group for the United Nations International Year of Rangelands and Pastoralists, 2026

Mobile pastoralism is a crucial livelihood for millions worldwide, supporting people and livestock across over half the world's land surface – the rangelands. This important book recognizes and supports this vital practice, which sustains communities in often harsh and hostile environments.

Jarso Mokku, CEO, Drylands Learning and Capacity Building Initiative, Nairobi, Kenya

Pastoralism, Uncertainty and Development

Edited by
Ian Scoones

Practical Action Publishing Ltd
25 Albert Street, Rugby,
Warwickshire, CV21 2SD, UK
www.practicalactionpublishing.com

© Ian Scoones and contributors, 2023

The moral right of the editor to be identified as editor of the work and the contributors to be identified as contributors of this work have been asserted under sections 77 and 78 of the Copyright Design and Patents Act 1988.

This open access publication is created under a Creative Commons Attribution Non-commercial No-derivatives CC BY-NC-ND licence.
This allows the reader to copy and redistribute the material, but appropriate credit must be given, the material must not be used for commercial purposes, and if the material is transformed or built upon the modified material may not be distributed. For further information see https://creativecommons.org/licenses/by-ncnd/4.0/legalcode

Product or corporate names may be trademarks or registered trademarks, and are used only for identification and explanation without intent to infringe.

A catalogue record for this book is available from the British Library.
A catalogue record for this book has been requested from the Library of Congress.

ISBN 978-1-78853-243-3 Paperback
ISBN 978-1-78853-244-0 Hardback
ISBN 978-1-78853-245-7 Electronic book

Citation: Scoones, I. (ed.), (2023) *Pastoralism, Uncertainty and Development*, Rugby, UK: Practical Action Publishing <http://doi.org/10.3362/9781788532457>.

Since 1974, Practical Action Publishing has published and disseminated books and information in support of international development work throughout the world. Practical Action Publishing is a trading name of Practical Action Publishing Ltd (Company Reg. No. 1159018), the wholly owned publishing company of Practical Action. Practical Action Publishing trades only in support of its parent charity objectives and any profits arecovenanted back to Practical Action (Charity Reg. No. 247257, Group VAT Registration No. 880 9924 76).

The views and opinions in this publication are those of the author and do not represent those of Practical Action Publishing Ltd or its parent charity Practical Action.

Reasonable efforts have been made to publish reliable data and information, but the authors and publisher cannot assume responsibility for the validity of all materials or for the consequences of their use.

Cover photo shows: Francesco Corda from northern Sardinia leading his sheep to graze.
Credit: Roopa Gogineni.
Typeset by vPrompt eServices, India

Contents

List of figures and tables	vii
List of abbreviations	ix
About the contributors	xi
Preface and acknowledgements	xv
Chapter 1: Pastoralism, uncertainty, and development: perspectives from the rangelands *Ian Scoones and Michele Nori*	1
Chapter 2: Decoding uncertainty in pastoral contexts through visual methods *Shibaji Bose and Roopa Gogineni*	21
Chapter 3: Engaging with uncertainties in the now: pastoralists' experiences of mobility in western India *Natasha Maru*	39
Chapter 4: Hybrid rangeland governance: ways of living with and from uncertainty in pastoral Amdo Tibet, China *Palden Tsering*	51
Chapter 5: Uncertainty, markets, and pastoralism in Sardinia, Italy *Giulia Simula*	65
Chapter 6: Responding to uncertainties in pastoral northern Kenya: the role of moral economies *Tahira Mohamed*	79
Chapter 7: Livestock insurance in southern Ethiopia: calculating risks, responding to uncertainties *Masresha Taye*	93
Chapter 8: Confronting uncertainties in southern Tunisia: the role of migration and collective resource management *Linda Pappagallo*	107
Chapter 9: Living with and from uncertainty: lessons from pastoralists for development *Ian Scoones and Michele Nori*	119
Index	141

List of figures and tables

Figures

1.1	Map of case study sites	2
2.1	A collection of Facebook groups dedicated to Douiret in southern Tunisia. Members post archival photos, maps, poetry, obituaries, and live videos of sheep-shearing and olive-picking.	23
2.2	Pastoralists fill jerry cans at the water pan in 1975 and 2020	24
2.3	A wealthy adult male pastoralist took this photograph to show a 'natural dinner' for children. Others perceived it differently.	25
2.4	An old photograph of a meal shared after sheep-shearing in Sardinia, Italy	26
2.5	Locust swarms in southern Ethiopia	27
2.6	Uncle Lhabe looks out at his former winter pasture, now underneath the lake	28
2.7	Goats lost to a lion	29
2.8	A Rabari camp on the move	31
2.9	An extended family gathers during a sheep-shearing in Douiret, Tunisia	32
2.10	Felice's cheese production, Sardinia	33
2.11	Tibetan language newspapers distributed at a horse festival	34
2.12	Swahili language newspapers on display at a community feedback session in Isiolo	35
2.13	Sardinian pastoralists looking at photos of Tunisian pastoralists	36
3.1	The morning after the hailstorm	40
3.2	Map of the Kachchh study area	43
4.1	Saga and Lumu in Amdo Tibet, China	52
4.2	Members of the local monastery carrying hay for the blue sheep in Golok	58
5.1	Map of Sardinia	70
5.2	Tonino with his Unifeed wagon. It simplifies distribution of feed, reduces labour costs, and allows for controlled and standardized animal feeding. Villamassargia, south-west Sardinia, November 2020.	72

5.3	Felice leading the flock to graze in the land he accesses through informal agreements with his neighbour. Sorso, north-west Sardinia, January 2020.	76
6.1	Map of the study area	82
6.2	'*Borantiti* [being Borana and showing the ideals of the Borana] is all about showing kindness and solidarity to overcome shortages. Here, women share labour to load water on the donkey.'	88
7.1	Study area: Borana, Ethiopia	97
7.2	A pastoralist fencing grazing land. When it starts raining, pastoralists enclose areas that are closer to their villages to improve grass growth for livestock feeding in the dry season.	101
8.1	Map of the study area	108
8.2	The old town of Douiret, with its troglodytic abodes, became largely unpopulated by the late 1980s as villagers moved to 'new' Douiret or elsewhere.	109
8.3	The *jessours* help with water and soil management in order to increase livelihood options	110

List of tables

4.1	Different responses to uncertainty in the Tibetan context	54
5.1	Distribution of sheep farms by flock size	66
7.1	Key features of the population from Gomole and Dire	98
7.2	Combining insurance with local responses in 2019 (per cent)	99
9.1	Reframing pastoral policy	131

List of abbreviations

CAP	Common Agricultural Policy
ESRC	UK Economic and Social Research Council
ERC	European Research Council
IBLI	Index-based livestock insurance
ILRI	International Livestock Research Institute
NDVI	Normalized Difference Vegetation Index
PASTRES	Pastoralism, Uncertainty and Resilience: Global Lessons from the Margins
TLU	Tropical Livestock Unit

About the contributors

Shibaji Bose is a creative consultant and a visual-methods researcher. His work draws on long-term visual ethnography and participatory visual action research in remote and climatically fragile zones in South Asia. He has written extensively on health systems and climate change in *The Lancet, Indian Anthropologist, BMC Health Services Research, BMJ Global Health, International Journal for Population, Development and Reproductive Health, Regional Environmental Change, The Statesman, The Economic Times,* and the *Hindustan Times*. Alongside his interests in mediating between dominant and implicit narrative spaces, he has co-curated photovoice exhibitions and directed films showcased by the Cannes Film Festival, Wellcome Trust, and Health Systems Global symposiums. He supported the photovoice work for the PASTRES (Pastoralism, Uncertainty and Resilience: Global Lessons from the Margins) programme.

Roopa Gogineni is a director and photographer focused on historical memory and modes of resistance. She has an MSc in African Studies from the University of Oxford, where she researched the construction of media narratives around Somalia. Her films have screened at festivals around the world including the International Documentary Film Festival Amsterdam, Hot Docs, Full Frame, and Sheffield DocFest. Her practice is rooted in long-term engagement and co-creation. Her short film *I Am Bisha* earned the Oscar-qualifying Full Frame Jury Award for Best Short, a One World Media Award, and a Rory Peck Award. She has directed documentaries for *The New York Times*, the BBC, and Al Jazeera and has received fellowships from CatchLight, Logan Nonfiction, and FRONTLINE/Firelight. She supported the visual methods work for the PASTRES programme.

Natasha Maru is a multidisciplinary social scientist and policy consultant working on pastoral development. She has recently finished a PhD with the PASTRES programme at the Institute of Development Studies, University of Sussex, UK, where she studied the temporal experiences of mobility among the Rabari pastoralists of western India. Drawing on a deep ethnography, she shows how pastoralists 'pace' themselves in a rapidly modernizing context. Her work innovatively combines concepts from mobility studies in sociology and geography with temporal studies and the long-running debates in pastoral development; it builds on earlier work undertaken for an MPhil in Development Studies at the University of Oxford. She is interested in space, the state, mobility, temporality, communing, and living life in technicolour.

Tahira Mohamed is an anthropologist from Marsabit County in northern Kenya. She recently completed her doctoral research under the PASTRES programme at the Institute of Development Studies, University of Sussex. Her study examines how 'moral economies' and social 'safety-net' institutions are evolving among the Waso Boran pastoralists of northern Kenya's Isiolo County. In particular, she looks at the everyday role of moral economy practices, including collective solidarities and resource redistribution in managing drylands variabilities such as drought, conflict, and other livelihood shocks since 1975. She holds a Masters' degree in International Studies from the University of Nairobi. Her MA project was on human smuggling across the Kenya–Ethiopia border. Tahira has worked on the politics of implementing social protection in Marsabit County as a field researcher with the Effective State and Inclusive Development Research Centre at the University of Manchester. Together with the Centre for Research and Development in Drylands, she is currently undertaking a study on local forms of resilience in pastoral areas, funded by the Australian Centre for International Agricultural Research.

Michele Nori is a tropical agronomist with a further specialization in rural sociology and specific expertise in the resource management and livelihood systems of agro-pastoral communities. Through 25 years of work experience, he has developed a 'horizontal career', by collaborating with various organizations including civil society, UN agencies, research institutes, agricultural enterprises, and donors' offices in different dryland regions. His publications range from scientific articles to technical notes and advocacy papers on agro-pastoral livelihoods. By integrating field practice and academic research, his concern is providing scientific evidence and policy advice on aspects of rural development, food security, and natural resource management. Such endeavours are currently undertaken through the Global Governance Programme of the Robert Schuman Centre at the European University Institute. He is part of the European Research Council (ERC) funded PASTRES programme, which aims to learn from pastoral systems means to tackle societal challenges related to growing uncertainties.

Linda Pappagallo has a background in research on the political economy of managing resources, which has focused on understanding the political economy of pastoral production, migration and collective management, particularly in the Mediterranean region. Her Economic and Social Research Council (ESRC) funded PhD research linked to the PASTRES *Partir pour Rester* (to leave in order to stay) programme explores how human mobility and collective herding practices in southern Tunisia explain the persistence of livestock-keeping. She focuses particularly on the role of 'absence' in changing agrarian settings and shows how pastoralism can persist as a livelihood through commoning practices. Linda is originally from Italy, although she grew up in different countries in the Middle East and North Africa. Before completing her PhD at the University of Sussex, she completed a Masters' in International Affairs at Columbia University. Linda is currently exploring practices of commoning in livestock-keeping, cheese-making and critical research on pastoral development through visual tools.

Ian Scoones is a professor at the Institute of Development Studies, University of Sussex. He is an agricultural ecologist by original training but today works on questions of policy around land, agriculture, and agrarian change, mostly in Africa. He is the principal investigator of the ERC-funded PASTRES programme (http://pastres.org). He was the co-director of the ESRC STEPS Centre from 2006 to 2021 and has been part of the editorial collective on the *Journal of Peasant Studies* over the last decade. He has published extensively on pastoralism and development since (co-)editing *Living with Uncertainty: New Directions in Pastoral Development in Africa* (IT Publications, 1995) and *Range Ecology at Disequilibrium: New Models of Natural Variability and Pastoral Adaptation in African Savannas* (ODI, 1993). His most recent publication is *Livestock, Climate and the Politics of Resources: A Primer* (TNI, 2022).

Giulia Simula is an agrarian and food movement researcher and activist originally from Sardinia, Italy. She currently works with the secretariat of the Civil Society and Indigenous Peoples' Mechanism for relations with the UN Committee of World Food Security. In her work, she facilitates the participation of small-scale food producers, pastoralists, peasants, the landless, and other sectors in policy processes. Giulia's PhD research with the PASTRES programme looked at the politics of pastoral markets, the dynamics of agrarian transformation, and the different ways in which pastoralists navigate market uncertainty in a globalized neoliberal economy. She supports the struggles of social movements and civil society organizations advocating for food sovereignty and for the right to adequate food and nutrition for all.

Masresha Taye is a development practitioner focusing on African dryland systems. He is interested in research, development, and policy work on drought and conflict-prone areas, disaster risk financing, anticipatory technologies/innovations for pastoral populations, and climate change and uncertainty. For five years, Masresha led the International Livestock Research Institute's (ILRI) index-based livestock insurance (IBLI) programme in Ethiopia. His PhD with the PASTRES programme explored how pastoralists combine different responses to drought with insurance. He is now a postdoctoral researcher at the University of Amsterdam (Amsterdam Institute for Social Science Research – Governance and Inclusive Development) where he is working on the project 'From Climate Change to Conflict: Mitigating through Insurance?'

Palden Tsering (Chinese Pinyin: Huadancairang) possesses an MSc from the Durrell Institute of Conservation Ecology University of Kent, UK and a PhD from the Institute of Development Studies, University of Sussex, UK, as part of the PASTRES programme. He has worked on the role of traditional

Tibetan resource governance, conservation and development, and the politics of these dynamic interactions amongst institutions (government, monastery, pastoral community) amid changes and uncertainties in the pastoralist context. His recent research is on hybrid rangeland governance in two pastoral settings in Amdo Tibet, China. His research offers a new way of thinking about land governance and suggests an approach to rangeland governance that goes beyond conventional approaches, with implications for management, policy, and the politics of land in the Tibetan–Chinese context.

Preface and acknowledgements

This book explores the connection between pastoralism, uncertainty, and development. It makes the case that recognizing how pastoralists make productive use of variability and embrace uncertainty is central to understanding how pastoral systems in marginal dryland and montane systems work. Further, the book argues that such understandings about how reliability is generated in the context of highly variable settings offer wider lessons for rethinking development policy and practice in today's uncertain, turbulent world.

This is important because uncertainties of all sorts – environmental, market-based, and political – are on the rise, as the world faces climate and environmental change as well as market volatility and political turmoil. Learning lessons from pastoralists therefore not only ensures that development efforts are more effective across the world's rangelands, where millions of pastoralists live, but is also important for all of us.

Pastoralists, while often marginalized in policy debates and development efforts, are important guardians of vast rangeland territories that make up over half the world's land surface. Pastoralism generates livelihoods for many, as well as providing animal-based products that enhance people's diets in some of the poorest parts of the world.

Despite their vital importance, pastoral systems are often deeply misunderstood, with false narratives dominating policy and public discourse alike. This book offers a different set of perspectives, rooted in in-depth research across six countries in three continents. The case studies presented as chapters of this book were developed as part of the PASTRES programme (Pastoralism, Uncertainty, Resilience: Global Lessons from the Margins, https://pastres.org) and individual PhD research projects over the past few years.

Together, they challenge mainstream thinking about pastoral development, offering a new narrative with variability and uncertainty at the centre. A unique lens on pastoralists' own understandings of variable and uncertain contexts is also offered through an innovative documentary photography and photovoice project. The photos have been exhibited across the world, including as part of feedback in all of the research sites, and in an online exhibition, Seeing Pastoralism (https://seeingpastoralism.org).

The book builds on long-term research on this theme, starting with the book *Living with Uncertainty: New Directions in Pastoral Development in Africa*, first published in 1995 by Intermediate Technology Publications and now available open access for the first time (https://practicalactionpublishing.com/book/1264/living-with-uncertainty). It also builds on the work conducted under the Future Agricultures Consortium pastoralism theme and published in two further books – *Pastoralism and Development in Africa: Dynamic Change at the Margins* (Catley et al. (eds), open access: https://doi.org/10.4324/9780203105979) and *The Politics of Land, Resources and Investment in Eastern Africa's Pastoral Drylands* (Lind et al. (eds), open access introduction, https://opendocs.ids.ac.uk/opendocs/handle/20.500.12413/15458) – published in 2012 and 2020 respectively.

The PASTRES programme has also produced a free online course (https://pastres.org/online-course/) and, together with the Transnational Institute, an accessible primer in multiple languages (https://www.tni.org/en/publication/livestock-climate-and-the-politics-of-resources), both of which pick up on many of the themes of this book. The PASTRES programme produces regular blogs and publishes a six-monthly newsletter, and these can be subscribed to via the website (https://pastres.org).

The PASTRES programme is supported by a ERC Advanced Grant (no. 70432) and is hosted by the Institute of Development Studies at the University of Sussex, UK and the European University Institute in Florence, Italy. The programme has involved six PhD students, four country lead researchers, five co-sponsored postdoctoral researchers, and a further 11 affiliate researchers; it has been led by Ian Scoones and Michele Nori, with support from Jeremy Lind and others. The communication and engagement activities of PASTRES have been led by Nathan Oxley, with support from Shibaji Bose, Roopa Gogineni, Ben Jackson, and Natasha Maru (https://pastres.org/about-us/pastres-team/).

This book is very much a collective output of the PASTRES team. The fabulous linocut illustrations were created by team member Linda Pappagallo, while the photographs in the book were curated by Roopa Gogineni. All the maps were prepared by John Hall. We would also like to thank our many allies and supporters across the world, as well as Andy Catley of Tufts University for reviewing the manuscript and Dee Scholey of Vital Editing and Catherine Fitzsimons for their excellent copyediting work. And, finally, many thanks to the team at Practical Action Publishing for their invaluable support during the publication process.

This book is a contribution to the wider efforts around the International Year of Rangelands and Pastoralists in 2026 (https://www.iyrp.info/).

Ian Scoones
Brighton, UK, January 2023

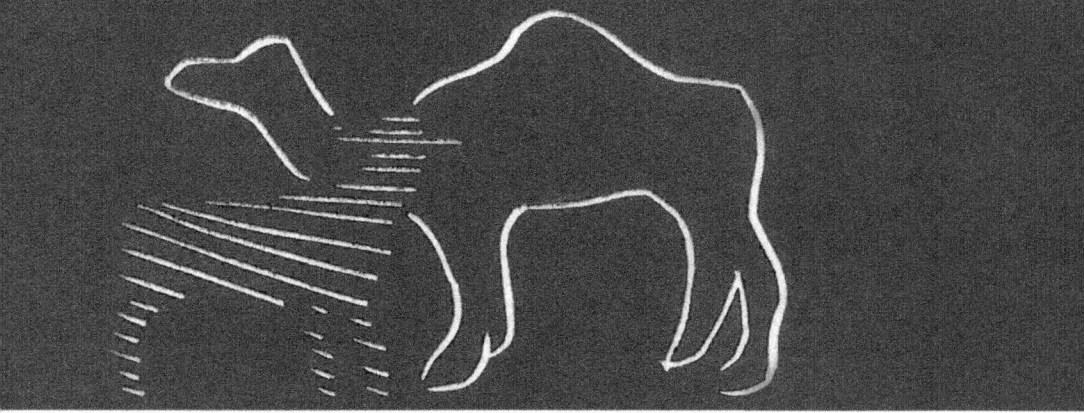

CHAPTER 1
Pastoralism, uncertainty, and development: perspectives from the rangelands

Ian Scoones and Michele Nori

Introduction

Uncertainties are everywhere in today's world. Market crashes, pandemics, climate change, war, and conflict all provide the backdrop to daily life. Uncertainty – where we don't know the likelihood of future events – must be central to any thinking about development (Stirling, 2010; Scoones, 2019; Scoones and Stirling, 2020). This is often forgotten in our push towards stable, predictable plans and our faith in risk-based approaches that claim the ability to predict the future through calculative models and technologies, offering early warnings to offset the worst. As recent events (not least the pandemic) have shown, this is not adequate, and uncertainty – even ignorance, where we don't know what we don't know – must be central to development thinking and practice.

Whom can we learn from in order to be better at responding to the uncertainties of our turbulent world? It is those who confront uncertainties on a daily basis and always have done so – pastoralists. Rooted in their cultural practices and institutions, pastoralists – like delta dwellers, shifting cultivators, or near-shore fishers – must live with and, indeed, from uncertainty (FAO, 2021; Krätli, 2015; Krätli and Schareika, 2010). In the drylands and mountains where pastoralists live, negotiating access to resources, navigating volatile markets, making use of varying social relations in times of stress, and responding to conflict and complex political dynamics are all essential if sustainable livelihoods are to be generated.

Yet the uncertainties that must be confronted are accelerating, emerging from wider structural changes in global political-economic relations including climate change, land grabbing, and the globalization of markets. Many of these

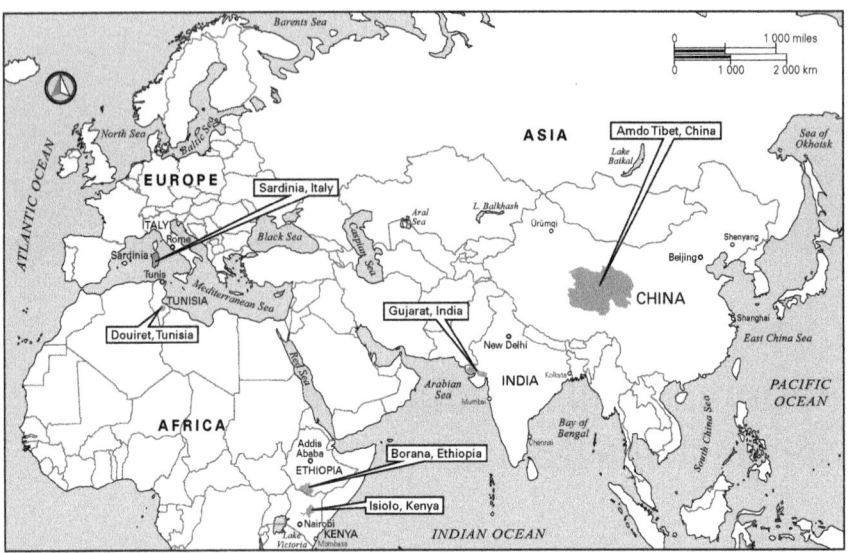

Figure 1.1 Map of case study sites

shifts are beyond the control of pastoralists themselves, making it essential for them to engage with others beyond their locales and organize collectively in order to respond to an increasingly turbulent world (Scoones 2022a).

This book argues that embracing uncertainty is essential for effective development. It also argues that pastoralists can help us reframe policies and practices in ways that go beyond a risk management and control approach to one that genuinely confronts situations where we don't know what the future holds. We consider this through a series of case studies, each focusing on a different theme and all emerging from the ERC-supported PASTRES programme (Pastoralism, Resilience and Uncertainty: Global Lessons from the Margins, http://pastres.org).

The cases come from three continents (Figure 1.1) – Africa (Ethiopia, Kenya, Tunisia), Asia (China and India), and Europe (Italy) – and encompass a range of themes, including mobility, resource governance, markets, moral economies, insurance, and communal institutions. The case studies are complemented by a chapter on visual, participatory methodologies as ways of exploring uncertainty through the eyes of pastoralists, and a final chapter that reflects on global and regional policy challenges. Overall, as this opening chapter outlines, the book offers a new perspective on pastoralism and development for our uncertain times.

Reframing development: what an uncertainty lens brings

Pastoralists must continuously confront uncertain events. Whether these are droughts, floods, or heavy snowfalls, they are a normal part of life and, at one level, are expected. Yet pastoralists never know when such events will occur, in what combinations, with what severity, over what area, and with what effects on pasture, disease ecologies, or markets. This is why flexible, adaptive practices are essential to live with and from uncertainty.

Conceptions of uncertainty are embedded in pastoralists' world-views and reflected in their practices (see Chapter 2). Pastoralists must respond in real time, learning and adapting along the way. Everyday practices, drawing on cultural norms and social relations, are central to responding to uncertainty, as the chapters that follow show. There are of course limits to such practices. As uncertainties increase – the result of wider, structural political-economic forces – the possibilities of local responses are inevitably constrained. In confronting uncertainty, a focus on the local must always be combined with attention to wider structural conditions and drivers.

A different perspective for development thinking and practice emerges from contrasting uncertainty (a lack of knowledge about future likelihoods of outcomes) with risk (where likelihoods can be calculated and predicted). This means a radical rethink of how we go about supporting people to manage resources and sustain livelihoods, adapt to climate change, negotiate markets, migrate to different places, deal with disasters, and so on. A new narrative for development, based on pastoralists' experiences, means thinking about how people – individuals but also, most importantly, collectives connected in networks – can transform high variability (the increasing norm) so as to ensure a reliable flow of goods and services (the desired outcome) (Roe, 2020).

However, much development policy and practice are blind to uncertainty. A framing around risk suggests the possibility of advance planning where stability and control are assumed. For some settings – such as building a bridge or road – a standard risk assessment is of course appropriate. There are detailed engineering protocols derived from the physics of materials, and deep understanding of potential shocks. But for contexts, we simply don't know what the future holds.

In respect of drought shocks, for example, even with the improvements in climate science, the models are too aggregated to know, beyond some very basic projections, what will happen in a particular place in a particular year. There are still other situations where there are 'unknown unknowns' – in other words, plain ignorance: where we know nothing about the outcomes or the likelihoods. These are surprise events, where things arrive out of the blue. These conditions of uncertainty and ignorance are common in complex, messy contexts – indeed, they are the norm in development settings, perhaps especially in pastoral areas.

How then are such conditions negotiated? One approach, developed in the context of studying critical infrastructures – water or energy supply systems, for example – highlights the vital role of 'high-reliability' professionals and networks in generating reliable livelihood options (Roe, 2013, 2016). Such high-reliability practices are rooted in local cultures and contexts but also rely on external support, technology, and information. High-reliability practices require scanning the horizon for troubles ahead while attending in real time to day-to-day responses. In supplying reliable flows of milk, meat, and other services, pastoralism can be seen as a critical infrastructure, populated by networks of high-reliability professionals taking on different roles in the system (Roe, 2020).

This is not an argument for a retreat to the local and total reliance on 'indigenous knowledge' – useful as such insights are – but an approach that gets us away from a control-oriented response that fails to engage with variability and the consequent uncertainties and sources of ignorance. A perspective that embraces uncertainty does not shy away from complexity and mess, domesticating challenges through calculative models and political technologies, but addresses uncertainties head-on.

Such a perspective must start from local conditions – the everyday experience of people on the ground. External support must articulate with this, helping rather than overwhelming local capacities. In pastoral areas, as the chapters show, well-meaning efforts – such as land governance reforms, insurance mechanisms, market support, and social protection programmes – will fail if they don't take uncertainty seriously. In the same way, even policies that recognize the importance of the mobility of pastoral peoples and livestock should avoid constraining this movement into narrow corridors or restricted time windows (see Chapter 9).

Responding to variability means temporal and spatial flexibility, with redundancy central to organizational design. Many examples are discussed in the chapters that follow, and the concluding chapter explores implications for policy, suggesting that a major reframing of policy narratives is needed for an approach that links pastoralism, uncertainty, and development. This means a shift from a commitment to 'control' – and prediction, stability, and planning – to one that is centred on social relationships and institutions that support flexible and adaptive responses to the inevitable uncertainties of today's world (cf. Scoones and Stirling, 2020).

What is pastoralism and why are rangelands important?

Pastoralism – the extensive use of rangelands through mobile livestock-keeping – is a vitally important livelihood practice globally. Rangelands cover more than half the world's land surface, supporting many millions of people, often in harsh and hostile environments (ILRI et al., 2021). The provision of livestock products – meat, milk, wool, hides, and so forth – is essential for local economies and the nutrition of often marginalized people (Manzano et al., 2021; Köhler-Rollefson, 2021).

Through highly skilled herding, often involving the movement of animals and people across time and space, pastoralists maximize production through exploiting environmental variability in places where other forms of food production are not feasible (Krätli, 2015; Nori, 2019a; FAO 2021). By managing extensive, biodiverse rangelands, pastoralists are also important for environmental protection and for generating food and income through mobile livestock systems (Köhler-Rollefson, 2023).

In this book, we look at pastoralism in the high-altitude mountains of Amdo Tibet in China, the Mediterranean hills of Sardinia in Italy, the savannas of East Africa in northern Kenya and southern Ethiopia, the dry plains of Kachchh in Gujarat in India, and the semi-desert and rocky mountains of southern Tunisia. We could have had examples from the tundra and forests of the Arctic, the grassland plains of the Americas or Australia, the temperate hills and mountains of northern Europe, and many more places. Pastoralists live in very diverse settings, under very different environmental conditions, influenced by diverse socio-economic and political contexts. What binds pastoralists together is their capacity to generate sustainable livelihoods from highly variable resources in a reliable way; and it is this capacity – and the core principles that emerge – which the cases in this book explore (Nori, 2019b; Scoones, 2021 and 2022a).

Rangelands are not uniform, despite the impression of a never-ending expanse of grassland with the odd tree. In some people's views, these are 'wastelands' in need of restoration through tree planting. But any pastoralist will tell you that rangelands are biodiverse open ecosystems (Bond, 2019). Scattered across the landscape are patches of different grass, multiple types of trees with diverse uses, and water sources, creating important wetter patches that act as key resources at the end of the dry season or during droughts (Scoones, 1991). With rainfall so variable year-on-year, many rangelands never reach an equilibrium carrying capacity – a fixed number of animals for an area of land. In one year there is plenty of grazing, but in the next year, there is virtually nothing and so it is vital to move to follow resource availability across the landscape. These are called non-equilibrium environments, where inter-annual fluctuations mean that conditions are never static and uncertainty is a defining feature (Vetter, 2005; Behnke et al., 1993; Scoones, 1995; Ellis and Swift, 1988).

All this means that standard approaches to resource management, premised on fixed, stable, unfluctuating conditions, do not work. Nor do standard designations of land degradation or desertification. It all depends on the baselines, and these shift radically and constantly. Attempts at control in highly variable, uncertain environments usually fail; fixed-tenure regimes, fencing, sedentarization schemes, ranching systems, delimited protected areas, and standardized stocking rates were designed for temperate settings, and do not translate to the high-variability contexts of most rangelands of the world. The illusion of control is strong and costly, but the alternatives, rooted in a more complex understanding of context, are often missed in these grand schemes. Such schemes are repeated again and again over decades of inappropriate interventions and failed development efforts, which frequently, undermine pastoralists' reliability capacities (de Haan, 1993).

Despite this, the mobile use of rangelands by pastoralists is under threat on a variety of fronts, with encroachment and fragmentation of rangelands continuing apace (Galvin et al., 2008; Lind et al., 2020a; Behnke, 2021). This increases uncertainty for pastoralists, as the land that is grabbed for agricultural or conservation investments, for example, usually includes the high-value, wetter patches which are essential for the functioning of the wider system. As a frontier for development, major investments are occurring across the world's rangelands, whether for irrigated agriculture, conservation areas, hunting and tourism, watershed management, mining, or alternative energy investments. All these efforts are combined with infrastructural investment – roads, rail lines, electricity and mobile phone connections, and more – often linked to the growth of towns in the peripheral pastoral areas. These wider processes of investment, extraction, and exclusion are having major impacts on pastoralists' abilities to live with and from uncertainty. Some of these investments benefit pastoralists – providing jobs, bringing services closer, and so on – but often increased investment on an unregulated frontier can lead to new forms of competition and to speculation, corruption, and deal-making that drive division within communities as elites become co-opted (Enns and Bersaglio, 2020; Lind et al., 2020a).

The challenges to pastoralism are exacerbated by the wider, longstanding, well-entrenched colonial narrative that pastoralists are 'backward', environmentally destructive, and in need of 'modernization' (Nori et al., 2008; Omondi and Odhiambo, 2009; Little, 2012; Nori, 2022). Seeing like someone from town, from the state, or from a development agency is very different to 'seeing like a pastoralist' from the variable, uncertain drylands (Catley et al. 2012; cf. Scott, 1998).

Pastoralism: a nature-positive contribution to the climate and biodiversity crises

Today, there is a strong Western and urban narrative about the dangers of livestock production for the climate and the wider environment. For example, aggregate statistics on the contributions of livestock production to climate change are deployed to argue for an end to livestock farming, with pastoralism and other systems wrapped up in a fervent rhetoric for a major switch in diets and production systems (Monbiot, 2022, but see Houzer and Scoones 2021). Of course, some industrial, contained systems of livestock production, reliant on imported feeds grown in cleared forested areas and transported across the world, are highly damaging to the environment. But failing to differentiate low-input, extensive livestock farming is a strategic error, with significant consequences for pastoralists the world over (García-Dory et al., 2021; Scoones, 2022b).

Even accounting for the limits of available data (which mostly come from high-intensity industrial systems), there is little doubt that pastoral production systems have low climate impacts and can, under the right conditions, have positive benefits for the environment. Pastoral animals are physiologically adapted to grazing conditions and are smaller than those in industrial systems; dung and urine is not concentrated in one place but spread around, potentially adding to carbon and nitrogen stores through incorporation by trampling, and grasslands can sequester and store carbon in large amounts. And in any case, low-density, dispersed pastoral systems may not add to emissions of greenhouse gases above the baseline of a 'natural', wildlife-base system (Manzano and White, 2019).

Pastoral production also produces incredibly valuable sources of nutrients – not just protein, but a whole array of critical nutrients, which are very difficult to gain from plant-sourced products (Leroy et al., 2022). They may also offer medicinal benefits such as those claimed for camel milk, for example. The option proposed by some, of producing meat alternatives by corporate, industrial in large fermentation vats or through lab-growing, seems to be a massive diversion, prone to corporate capture, and forgetting the livelihood and environmental benefits of pastoral livestock production (Howard, 2022).

Another argument often laid at the door of pastoralists is that they are environmentally destructive, and their use of fragile, biodiverse environments should be curtailed in favour of protective, exclusionary conservation and forms of rewilding. This again has a long history, dating back to colonial views of rangelands, where the lack of trees was seen as an indicator of deforestation, with many rangelands designated 'wastelands' where 'desertification' was rife (Behnke and Mortimore, 2016; Davis, 2016). However, this narrative derives from a basic misunderstanding of the dynamics of open ecosystems and the importance of variability in rangelands (Bond, 2019; Vetter, 2020).

Despite much debate on these issues over many decades, the control imperative of environmental efforts remains, reinforced through inappropriate interventions. Huge tree-planting campaigns are proposed for the world's rangelands, with vast targets for expanding protected areas to preserve biodiversity. In some parts of the world, largely urban movements are arguing for 'rewilding' where livestock and people are removed (or drastically reduced) and an alternative ecosystem is imagined, with trees regrowing and wild predators reintroduced.

Many of these schemes are again premised on a false understanding of rangelands as sub-climax forests, rather than natural systems maintained by a combination of grazing and fire over millennia, with or without human intervention. Of course, concentrating animals and people and reducing movement options – as is happening through the effect of enclosures from investments and land and green grabs – will result in land degradation, but this is not the inherent consequence of a pastoral system of production. On the contrary, mobile pastoralism can maintain, even enhance, biodiversity through patch-based grazing, the dispersed deposition of manure and urine creating nutrient hotspots, the spreading of seeds through movement along transhumant routes connecting habitats, and the synergistic interaction with both wild predators and scavengers, which are often endangered species in such environments.[1]

In sum, pastoralists – as producers who can make use of variability in non-equilibrium rangelands and who know how to confront uncertainty in harsh, unstable ecosystems – can be good for the environment. They can have low impacts on the climate, sometimes even positive ones through the sequestration in the soil of carbon and nitrogen, and positive impacts on biodiversity, preserving open ecosystems through the skilled management of grazing and fire. And through such a well-adapted production system, pastoralists generate livelihoods and support the wider economy through many actors: service providers, market brokers, transporters, and others who process and trade animal products. By providing nutrient-dense, high-quality food, pastoral production can help reduce malnutrition and improve food security in some of the poorest parts of the world, while selling on to urban consumers elsewhere a potentially climate- and biodiversity-friendly product for which there is growing demand (Manzano et al., 2021; Köhler-Rollefson, 2021).

Challenges for pastoral development

This is not to say that all is well in the pastoral rangelands of the world. As the case study chapters show, there are many challenges. The adaptive flexibility at the heart of pastoralists' responses to variability and uncertainty may not always work. Strategies developed decades ago may not be sufficient to sustain fast-growing populations and may be unable to confront the more frequent droughts, floods, and compounding uncertainties faced today. Pastoralists must always innovate, adapt, and change to new circumstances. However, things are not always easy as a result of constrained access to resources, terms of trade that penalize pastoral production, and state or donor support that is often lacking or misplaced, given the false narratives that still dominate policy thinking (see Chapter 9).

As challenges have increased, in many places there has been a growing differentiation of pastoralists between rich and poor, urban- and rural-based, those with other jobs and connections and those without, those farming and those who do not, and so on. Many younger people from pastoral families are reluctant to follow their parents into what are seen as harsh, unrewarding pastoral livelihoods and seek opportunities elsewhere through migration. There are thus emerging class divisions within pastoral settings that can result in conflict and exploitation (Catley et al., 2012; Scoones 2021). Some richer households become increasingly oriented towards commercial livestock production, investing in technologies to intensify production, hiring in herders, enclosing rangelands, and capturing reliable markets. In some instances, extensive livestock production becomes the speculative activity of rich and well-connected entrepreneurs, traders, and politicians, sometimes from outside the area.

These changes leave others to struggle with fewer resources and limited market access. When major shocks arise – a drought, for example – some drop out, as their limited herds and flocks are depleted and they must make a living outside pastoralism or provide services or labour to the richer pastoralists. Some may become destitute and migrate to other areas or seek help from humanitarian agencies. While, over time, some may return, it is becoming more and more difficult to do so (Lind et al., 2020b). Impoverishment, socio-economic inequality, political marginalization, and a sense of grievance may be the triggers for conflict and insecurity, resulting in resource competition, increasing ethnic and cultural divides linked to forms of political capture, and radicalization (Benjaminsen and Ba, 2019; Nori, 2022).

This book does not attempt to paint a rosy picture of an imagined, pastoralist idyll now long-lost – if it ever even existed. The case studies discussed across the chapters and introduced next provide a flavour of the complex, contested, and highly differentiated realities in the different sites, influenced by diverse political economies. What comes across, however, is that, despite the challenges, pastoralists are continuously adapting to contexts of high variability and uncertainty and, as we discuss in Chapter 9, offer important lessons for us all.

The case studies

The cases discussed in the following chapters highlight how forms of variability, and therefore uncertainty, are changing and how environmental, market, social, and political factors combine in different ways. The wide array of practices that pastoralists deploy is discussed, showing how reliability and resilience are generated in different settings in response to diverse shocks and stresses, very often exploiting variability as a productive resource, rather than a threat. This section introduces the cases and sets them within the wider story of pastoralism in their respective regions. In Chapter 9, we return to these regional settings and explore the implications for development policy and practice.

Kachchh, Gujarat, India

Kachchh is a huge dryland area in the west of India, bordering Pakistan, with a mix of extensive grasslands and increasingly cropped areas. As elsewhere, rainfall variability is growing, and in recent years, the grasslands have been subject to flooding. The area hosts a mix of pastoralists from a number of different groups, including the Rabari. As with the rest of India, the Green Revolution has affected Gujarat through the expansion of intensive agriculture. New irrigation infrastructure and policies in favour of commercial agriculture support this expansion.

In 2001, Kachchh experienced a massive earthquake and there was a subsequent drive to industrialization as part of rebuilding the region, again transforming the landscape into one of fragmented and diversified uses. Today, pastoralists have to navigate between different land uses, moving between grazing areas using different forms of transport, and adapting herding strategies accordingly. Links to farming are crucial, as pastoralists negotiate access to crop residues on farmers' fields. Such relations are essential in sustaining pastoralism in the area by providing access to rich, nutritionally diverse fodder resources. Each year, the pattern of movement is different, responding to the timing of rains, crop harvests, and animal breeding, among other factors. Movements follow different paths but link to cultural and religious sites and connect grasslands to croplands.

As Chapter 3 explains, the contingent performance of movement is invested with social, cultural, and emotional meaning for the Rabari pastoralists and cannot be explained simply as a 'rational' response to variable conditions. The rhythm and pace of movement are of course conditioned by environmental dynamics and so generate reliability in the face of uncertainty, but such responses are bound up with social and political dynamics within mobile groups, between pastoralists, and across pastoralists, farmers, and government officials.

Amdo Tibet, China

Amdo Tibet stretches across the Qinghai plateau from the Kokonor lakeshore to the high mountains of Golok. High-altitude pastoralism is practised across this area, with the herding of yaks, sheep, and goats in particular. Pastoral production involves movement between winter pastures in the lowlands and summer pastures in the mountains. Skilled herding allows the management of grazing across these sites. While people live in villages near the winter pastures, they move with tents to the high pastures in the summer. In some places, herding competes with the harvesting of caterpillar fungus, a valuable medicinal product that grows in certain pastures.

Amdo Tibet has been subject to major changes in the past years, as China modernizes its economy and increases its influence in the Tibetan areas. Investments in national parks, linked to watershed protection for China's major river systems, have removed large areas from pastoral use. Large infrastructure developments associated with China's Belt and Road Initiative have seen roads, rail lines, and major energy schemes being built in the pastoral areas. Various tenure reforms have attempted to individualize range ownership, with particular plots assigned to certain households. Quota systems, regulated at village level, allow certain numbers of animals in defined areas.

As part of Chinese development programmes to improve living conditions and access to primary services for those living in under-developed provinces, the state has developed towns in pastoral areas, including resettlement housing for pastoralists. For some, this is an imposition that undermines herding systems while, for others, it provides welcome services. Herders split their households, with some staying with the animals in the mountains while others stay in the resettlement schemes. Negotiating this new landscape, in which there is an increasing presence of the state as well as in-migration from other areas of the country, means pastoralists must draw on diverse relationships, with a view to exploit ecological and institutional opportunities.

Gaining access to land for grazing is less straightforward than before and the individualized plot system does not allow the flexibility to manage grazing effectively. As a result, hybrid systems of rangeland governance have evolved (see Chapter 4). Such systems are neither private, nor communal, nor completely open property arrangements: they emerge from a negotiation between herd-owners, village heads, religious monastery leaders, government officials, and others. Institutional and organizational innovation therefore means that pastoralists can generate reliability in the face of new forms of variability and uncertainty.

Sardinia, Italy

Sardinia, an island in the Mediterranean off mainland Italy, has the largest population of sheep in Europe. It also produces the largest quantity of pecorino cheese in the world, most of it exported to Europe and the USA. Milk production from sheep is therefore central to the livelihoods of many Sardinians and keeping sheep has long been central to the culture and economy of the island. Indeed, many Sardinians migrated as herders to mainland Italy in the nineteenth century. However, in Sardinia, there are many styles of sheep-milk production, associated with different types of environmental, institutional, and market engagement.

There is no one single type of Sardinian pastoralist. They include those who have intensified heavily with investment in machinery for milking, in producing high-value fodder crops, and in supplying their milk to industrial cheese processors. Such livestock production does not respond flexibly to variability through herding but creates stability through technological, managerial, and contractual arrangements. By contrast, other pastoralists are more reliant on open pasture, where herding between different sites is important.

Although the traditional transhumance from the mountains to the lowlands has declined, flocks are still on the move locally, with pastoralists always aiming to manage variability so as to produce reliably. Such pastoralists range from those who sell milk to cooperative dairies, to those who have invested in mini-dairies to produce artisanal products themselves, to those who just sell locally and informally, focusing mostly on home consumption and tourist markets. Expanded social networks, changing agricultural policies, and diversified market options have therefore reconfigured the landscape of Sardinian pastoralism, with new uncertainties emerging.

As Chapter 5 explains, understanding how uncertainties – in the weather, in the market, in political conditions – are confronted by pastoralists requires insights into how markets and livelihoods are connected. Changing market conditions require adaptation of livelihood practices, including the management of flocks. Equally, when production conditions change – say, through a poor season – market strategies must be changed too. Thus, understanding how reliability is generated in pastoral systems requires an intimate knowledge of markets, how they are constructed socially and politically, and how they change over time.

Isiolo, northern Kenya

The drylands of northern Kenya are home to many pastoral groups, but Isiolo County is dominated by the Borana. Traditionally cattle-keepers, they are increasingly switching to camels and goats due to climate and other challenges, which implies a significant reorganization of herding practices. Within Isiolo, there are varied rangeland areas, criss-crossed by seasonal rivers which provide important dry-season grazing for animals. Recent decades have seen a dramatic increase in livestock trade, including for export, and more recently a huge growth in camel-milk marketing. The expansion of these trade systems has contributed to the development of diverse market networks, supported by newurban hubs and road connections. These in turn generate new opportunities for pastoral producers, women traders, truck transporters, and motorbike riders, all supporting an expanding pastoral economy.

Surrounded by Somali and Samburu areas, as well as increasingly policed national parks and conservancies, pastoralists in Isiolo feel hemmed in and are suffering an increasing fragmentation of land. This is the result of growing encroachment, as well as land speculation emerging from the development of transport corridors, wind farm development, and oil exploration. The result is heightened conflict: between ethnic groups seeking out scarce grazing; between humans, livestock, and wildlife; and with investors and land speculators of different sorts. With the increase in small arms in the region, some of these conflicts are violent, resulting in areas becoming out of bounds for grazing by animals. This adds to the uncertainties around gaining access to resources, requiring Boran pastoralists to develop new ways of coping.

For some, links to growing urban centres are essential and some pastoralists have settled, hiring herders to manage their animals in the bush. Others have moved out of pastoral production altogether and are involved in the wider service economy, often providing support to pastoralists. Responding to different forms of uncertainty has always been part and parcel of pastoralists' livelihoods, but the uncertainties, and their frequency and intensity, have changed. Climate change, land fragmentation, market volatility, disease outbreaks, and conflicts compound each other. The result is that pastoralists must rely on both individual and collective ways of responding to uncertainty and reducing the impacts of high levels of variability to assure livelihoods.

Traditionally, Boran society had many different forms of sharing and collective organization to manage herds, grazing, labour, and marketing. Chapter 6 explores the persistence of local forms of 'moral economy' centred on redistributive practices and collective solidarity across family, village, and clan. Reliability in the face of shocks and stresses is generated through social relations and networks, reinforced by cultural norms and local institutions, both new and old.

Borana, southern Ethiopia

Across the border to the north, Borana is a wide zone with a large pastoral population. The area is wetter than many other pastoral areas in the region and, over the last 30 years, agriculture has expanded massively, particularly in the areas around Yabello, the main town in the region. Here, pastoralists combine crop-farming with livestock-keeping (mostly cattle but also smallstock) and engage extensively in trade and other activities linked to the now large urban centre. Further south, where the rangelands are less connected and drier, a more traditional extensive pastoralism is practised. High levels of rainfall variability, a strict policy framework, increasing encroachment through farming, and conflict with neighbouring groups have meant that pastoral production is challenging. Access to grazing is dwindling, long-distance movements to other areas during drought periods are difficult because of insecurity, and there are also the challenges of mobilizing labour and capital.

The traditional network of collectively managed wells continues to operate, but pressure on water resources has increased as privatization through the drilling of boreholes has affected the management of both grazing and water. As conditions change, some pastoralists have moved out of livestock production, taking up jobs in local towns or migrating further away. Meanwhile, other pastoralists have been able to accumulate by taking advantage of growing milk commercialization and export markets for cattle, and the opportunities for selling on to traders who fatten animals in the highlands. The result is a much more differentiated pastoral population. While kept together by social commitments to the clan and wider Boran identity, processes of social differentiation – accumulation for some, increasing poverty or destitution for others – is evident. This has both a gender and generational dimension, with women and young people in particular losing out.

Responding to the challenge of drought in pastoral areas, numerous projects have been initiated aimed at increasing the resilience of livelihoods and assuring social protection. Many of these have failed, and agencies have been looking for new approaches compatible with pastoral settings. The latest effort, index-based livestock insurance (IBLI), has become significant over the last decade, as Chapter 7 explains. This aims to provide payouts to pastoralists in advance of a disaster taking hold, based on an index linked to pasture conditions assessed by satellite imagery. The insurance product relies on a risk assessment of a single peril – a lack of rainfall and therefore pasture – and does not take account of the complex combination of uncertainties that pastoralists must address. A focus on singular, calculable, index-based risk rather than multiple, intersecting uncertainties means that, for many, the insurance product is not appropriate or must be combined with other responses within their wider livelihood portfolio.

Tataouine, southern Tunisia

At the crossroads between the Sahara, the Maghreb, and Europe, the dry desert margins of southern Tunisia are harsh areas, with low and variable rainfall, where integrated forms of resource management such as the terraced *jessours* offer valuable grazing to the Amazigh, Berber communities inhabiting Douiret in the mountainous ranges south of Tataouine. Here, livestock production is combined with agriculture, mostly olive growing; however, making a livelihood in the mountains is tough and, increasingly, livelihoods in Douiret are crafted with income from elsewhere. As chapter 8 discusses, these areas have a long migration history connecting Douiret to other parts of Tunisia, notably the capital Tunis, but also to France, Canada, and parts of the Arabian Gulf. The Douiri diaspora is huge but still remains culturally and economically connected to the pastoral areas, as remittances from outside the area provide the basis for investment and accumulation in local flocks.

Such accumulation processes are not linear or even. Livestock exists as 'liquid wealth', subject to boom-and-bust cycles, and flocks grow and decline, not only in relation to environmental conditions but also due to the flow of remittances from outside. Since migrants are absent, either seasonally or very often for long stretches at certain stages of people's lifecycles, ways of managing livestock in their absence are required. Sometimes, this falls on women and the young and old who live in the villages more consistently, but such support is often combined with collective arrangements for managing flocks, including hired herding, sharing pasturelands, and collaborative tending of animals. Such arrangements are organized around collective pooling systems, such as the *khlata,* which also offer a link between migrants and their flocks at home, assuring their management and production while they are absent.

The process of accumulation of livestock in this setting is thus mediated by family arrangements and collective institutions, and is dependent on both variabilities within the rangelands and variabilities of income earnings outside and inside Douiret. The evolution of social networks is also important for gaining access to market opportunities and political support, both of which contribute to reconfiguring local pastoralism. The result is an uneven, non-linear pattern of flock growth, reflecting economic, social, and political dynamics across locations. This generates a particular pastoral mode of accumulation and redistribution/sharing, conditioned both by the nature of livestock as capital and by the types of institutions that mediate how flocks grow.

Characterizing pastoral systems

Across these settings, a number of principles emerge that characterize pastoral systems (see Krätli et al., 2015; Nori, 2019b; Scoones, 2021):

- First is the nature of livestock management, the intimate human–animal connections that allow variability to be exploited, ensuring that herds and flocks grow and produce, and that disease or death are avoided. This requires skilled herding, including the switching of species, composition, and careful breeding and training to encourage certain traits and behaviours.
- Second is the importance of livelihood diversification. In pastoral areas, there is always a complex mosaic of livelihood strategies, combining livestock-keeping with other activities including farming, trading, service provision, and so on. Increasingly, there are urban connections, sometimes linked to long-term migration and the reconfiguration of gender and generational roles. In order to gain value from livestock for livelihoods, engaging with diverse markets – always embedded in social and political relations and livelihood contexts – is essential.
- Third, mobility is essential for all pastoral production – whether of animals, fodder, water, or marketed products. This is a key response to the variability of rangeland systems and essential for survival. With increasing rangeland fragmentation due to encroachments of different sorts, patterns of mobility must change. Movement of people is important too, with migration to and from other places becoming the norm, and migrants providing remittances or hired labour, for example. Such connections over space and time are crucial in responses to uncertainty, with collective institutions emerging in pastoral settings to allow for those absent to continue to engage with pastoralism.
- Fourth, responses to uncertainties often require close connections with agriculture to be formed, either as part of a wider set of enterprises pursued by an agro-pastoral household or through striking up relations with farmers to gain access to fodder. Pastoral territories are not the unenclosed vast pastures that are sometimes imagined but are increasingly divided, with different patches enclosed and privatized and affected by wider processes of investment, intensification, and accumulation by the state and both local and global investors. This requires new forms of land control and management across reticulated territories, with flexible, negotiated forms of tenure and hybrid land uses being essential.
- Fifth, in order to be able to manage land, markets, and social and political relations with neighbours, the state, and other 'outsiders', pastoralists must be increasingly adept at developing relationships, building networks, and negotiating access to resources. Investment in social institutions is vitally important as a reliance on such communal, redistributive institutions is essential for offsetting the impacts of

high variability and uncertainty and generating reliable, sustainable livelihoods in increasingly challenging circumstances.

Wider structural political economy forces may, however, undermine pastoralists' strategies for generating reliability in the face of variable conditions, as the chapters discuss. In some contexts, the constraints may simply be too great: there may not be enough land if it is degraded, grabbed, and enclosed; climate variability may be too extreme for local strategies to work; the terms of trade in markets may be too skewed for pastoralists to make a living; and so on. Understanding how uncertainties are constructed and responded to within wider political economies, and how global and national forces impinge on pastoral settings, is therefore vital if the connections between pastoralism, uncertainty, and development are to be fully understood and appropriate interventions designed. Just like everyone else, pastoralists cannot be simply expected to go it alone.

Conclusion

As the chapters in this book show, pastoralism is a modern and intensive production system, adapted to highly variable contexts and able to adjust to fast-changing uncertainties in an increasingly turbulent world – although of course with limits. Pastoralism therefore remains a viable, vibrant production system precisely because pastoralists are able to live with and from uncertainty, transforming high variability into successful livelihoods, even under very challenging circumstances.

The sources and impacts of uncertainty change over time and across contexts, as the book's case studies show. Today, climate change, expanding investments and infrastructure, and engagement in global markets all affect pastoral systems, generating new uncertainties. For pastoralism to persist, pastoralists must continuously adapt and innovate. The result is that 'traditional' practices – transhumance, collective wells, land institutions, livestock sharing and loaning, and so on – must always change; new practices, technologies, social institutions, mobility patterns, and networked relations are always emerging.

Understanding the different dimensions of pastoralism and pastoralists' adaptive capacities can help us learn wider global lessons on how to respond to diverse uncertainties. At the root of pastoralism's success – in all its diversity – are the set of principles centred on making use of variability and managing uncertainty highlighted above and discussed throughout the chapters that follow. These principles, with their focus on the key attributes of flexibility, adaptation, innovation, and learning for generating reliability, are in turn suggestive of a wider set of principles for development more generally. This book therefore offers broader lessons for development and governance the world over that go beyond the standard, rigid modes of risk management, planning, and control. These lessons are developed further in Chapter 9.

Note

1. See the set of PASTRES briefings, www.pastres.org/biodiversity

References

Behnke, R.H. (2021) 'Grazing into the Anthropocene *or* back to the future?' *Frontiers in Sustainable Food Systems* 5: 638806 <https://doi.org/10.3389/fsufs.2021.638806>.

Behnke, R. and Mortimore, M. (eds) (2016) *The End of Desertification? Disputing Environmental Change in the Drylands*, Springer, Berlin.

Behnke, R.H., Scoones, I. and Kerven, C. (eds) (1993) *Range Ecology at Disequilibrium: New Models of Natural Variability and Pastoral Adaptation in African Savannas*, Overseas Development Institute, London.

Benjaminsen, Tor A. and Ba, B. (2019) 'Why do pastoralists in Mali join jihadist groups? A political ecological explanation', *Journal of Peasant Studies* 46: 1–20 <https://doi.org/10.1080/03066150.2018.1474457>.

Bond, W.J. (2019) *Open Ecosystems: Ecology and Evolution Beyond the Forest Edge*, Oxford University Press, Oxford <https://doi.org/10.1093/oso/9780198812456.001.0001>.

Catley, A., Lind, J. and Scoones, I. (eds) (2012) *Pastoralism and Development in Africa: Dynamic Change at the Margins*, Routledge, London <https://doi.org/10.4324/9780203105979>.

Davis, D.K. (2016) *The Arid Lands: History, Power, Knowledge*, MIT Press, Cambridge MA.

Ellis, J.E. and Swift, D.M. (1988) 'Stability of African pastoral ecosystems: alternate paradigms and implications for development', *Journal of Range Management* 41: 450–59 <https://doi.org/10.2307/3899515>.

Enns, C. and Bersaglio, B. (2020) 'On the coloniality of "new" mega-infrastructure projects in East Africa', *Antipode* 52: 101–23 <https://doi.org/10.1111/anti.12582>.

FAO (2021) *Pastoralism – Making Variability Work*, FAO Animal Production and Health Paper 185, Rome <https://doi.org/10.4060/cb5855en>.

Galvin, K.A, Reid, R.S., Behnke, R.H. and Hobbs, N.T. (eds) (2008) *Fragmentation in Semi-Arid and Arid Landscapes: Consequences for Human and Natural Systems*, Springer, Dordrecht.

García-Dory, F., Houzer, E. and Scoones, I. (2021) 'Livestock and climate justice: challenging mainstream policy narratives', *IDS Bulletin* [online first] <https://opendocs.ids.ac.uk/opendocs/handle/20.500.12413/16913>.

Haan, C. de (1993) *An Overview of the World Bank's Involvement in Pastoral Development*, paper presented at the Donor Consultation Meeting on Pastoral National Resource Management and Pastoral Policies for Africa organized by the United Nations Sudano-Sahelian Office, Paris, December 1993.

Houzer, E. and Scoones, I. (2021) *Are Livestock Always Bad for the Planet? Rethinking the Protein Transition and Climate Change Debate*, PASTRES, Brighton <https://doi.org/10.19088/STEPS.2021.003>.

Howard, P. (2022) 'Cellular agriculture will reinforce power asymmetries in food systems', *Nature Food* 3: 798–800 <https://doi.org/10.1038/s43016-022-00609-5>.

ILRI, IUCN, UNEP and ILC (2021) *Rangelands Atlas*, International Livestock Research Institute, Nairobi.

Köhler-Rollefson, I. (2021) *Livestock for a Small Planet*, League for Pastoral Peoples and Endogenous Livestock Development, Ober Ramstadt. Available from: <http://www.ilse-koehler-rollefson.com/wp-content/uploads/2021/10/livestock-for-a-small-planet_web.pdf>.

Köhler-Rollefson, I. (2023) *Hoofprints on the Land: How Traditional Herding and Grazing Can Restore the Soil and Bring Agriculture Back in Balance with the Earth*, Chelsea Green, London.

Krätli, S. (2015) *Valuing Variability: New Perspectives on Climate Resilient Drylands Development*, International Institute for Environment and Development, London. Available from: <http://pubs.iied.org/10128IIED.html>.

Krätli, S. and Schareika, N. (2010) 'Living *off* uncertainty: the intelligent animal production of dryland pastoralists', *The European Journal of Development Research* 22: 605–22 <https://doi.org/10.1057/ejdr.2010.41>.

Krätli, S. et al. (2015) *A House Full of Trap Doors: Identifying Barriers to Resilient Drylands in the Toolbox of Pastoral Development*, IIED Discussion Paper, London and Edinburgh. Available from: <http://pubs.iied.org/10112IIED>.

Leroy, F. et al. (2022) 'Animal board invited review: animal source foods in healthy, sustainable, and ethical diets – an argument against drastic limitation of livestock in the food system', *Animal* 16: 100457 <https://doi.org/10.1016/j.animal.2022.100457>.

Lind, J., Okenwa, D. and Scoones, I. (2020a) 'The politics of land, resources & investment in Eastern Africa's pastoral drylands', in J. Lind, D. Okenwa and I. Scoones (eds), *Land Investment & Politics: Reconfiguring Eastern Africa's Pastoral Drylands*, pp. 1–32, James Currey, Woodbridge.

Lind, J., Sabates-Wheeler, R., Caravani, M., Biong Deng Kuol, L. and Manzolillo Nightingale, D. (2020b) 'Newly evolving pastoral and post-pastoral rangelands of Eastern Africa', *Pastoralism* 10: 1–14 <https://doi.org/10.1186/s13570-020-00179-w>.

Little, P. (2012) 'Reflections on the future of pastoralism in the Horn of Africa', in A. Catley, J. Lind and I. Scoones (eds), *Pastoralism and Development in Africa: Dynamic Change at the Margins*, pp. 243–49, Routledge, London.

Manzano, P. and White, S.R. (2019) 'Intensifying pastoralism may not reduce greenhouse gas emissions: wildlife-dominated landscape scenarios as a baseline in life-cycle analysis', *Climate Research* 77: 91–7 <https://doi.org/10.3354/cr01555>.

Manzano, P. et al. (2021) 'Toward a holistic understanding of pastoralism', *One Earth* 4: 651–65 <https://doi.org/10.1016/j.oneear.2021.04.012>.

Monbiot, G. (2022) *Regenesis: Feeding the World without Devouring the Planet*, Penguin, New York NY.

Nori, M. (2019a) *Herding through Uncertainties – Regional Perspectives: Exploring the Interfaces between Pastoralists and Uncertainty: Results from a Literature Review*, Working Paper 68, Global Governance Programme, EUI Robert Schuman Centre, Firenze. Available from: <https://cadmus.eui.eu/handle/1814/64165>.

Nori, M. (2019b) *Herding through Uncertainties – Principles and Practices: Exploring the Interfaces between Pastoralists and Uncertainty: Results from a*

Literature Review, Working Paper 69, Global Governance Programme, EUI Robert Schuman Centre, Firenze. Available from: <https://cadmus.eui.eu/handle/1814/64228>.

Nori, M. (2022) *Assessing the Policy Frame in Pastoral Areas of Sub-Saharan Africa (SSA)*, Policy Paper 2022/03, Global Governance Programme, EUI Robert Schuman Centre, Firenze. Available from: <http://hdl.handle.net/1814/74314>.

Nori, M., Taylor, M. and Sensi, A. (2008) *Browsing on Fences: Pastoral Land Rights, Livelihoods and Adaptation to Climate Change*, IIED Drylands Issues Paper 148, International Institute for Environment and Development, London. Available from: <http://www.iied.org/pubs/display.php?o=12543IIED>.

Omondi, S.S. and Odhiambo, M.O. (2009) *Pastoralism, Policies and Practice in the Horn and East Africa: A Review of Current Trends*, Humanitarian Policy Group Commissioned Report, Overseas Development Institute, London.

Roe, E. (2013) *Making the Most of Mess: Reliability and Policy in Today's Management Challenges*, Duke University Press, Durham NC.

Roe, E. (2016) 'Policy messes and their management', *Policy Sciences* 49: 351–72.

Roe, E. (2020) 'A new policy narrative for pastoralism? Pastoralists as reliability professionals and pastoralist systems as infrastructure', *STEPS Working Paper* 113, STEPS Centre, Brighton. Available from: <https://opendocs.ids.ac.uk/opendocs/bitstream/handle/20.500.12413/14978/STEPS_WP_113_Roe_FINAL.pdf?sequence=105&isAllowed=y>.

Scoones, I. (1991) 'Wetlands in drylands: key resources for agricultural and pastoral production in Africa', *Ambio* 20: 366–71.

Scoones, I. (ed.) (1995) *Living with Uncertainty: New Directions in Pastoral Development in Africa*, Intermediate Technology Publications, London. Available from: <https://practicalactionpublishing.com/book/1264/living-with-uncertainty>.

Scoones, I. (2019) 'What is uncertainty and why does it matter?', *STEPS Working Paper* 105, STEPS Centre, Brighton. Available from: <https://opendocs.ids.ac.uk/opendocs/bitstream/handle/20.500.12413/14470/STEPSWP5_Scoones_final.pdf?sequence=1&isAllowed=y>.

Scoones, I. (2021) 'Pastoralists and peasants: perspectives on agrarian change', *Journal of Peasant Studies* 48: 1–47 <https://doi.org/10.1080/03066150.2020.1802249>.

Scoones, I. (2022a) *Livestock, Climate and the Politics of Resources: A Primer*, Transnational Institute, Amsterdam. Available from: <https://www.tni.org/en/publication/livestock-climate-and-the-politics-of-resources>.

Scoones, I. (2022b) 'Livestock, methane and climate change: the politics of global assessments', *WIREs Climate Change* 2022:e790 <https://doi.org/10.1002/wcc.790>.

Scoones, I. and Stirling, A. (eds) (2020) *The Politics of Uncertainty: Challenges of Transformation*, Routledge, London. Available from: <https://library.oapen.org/handle/20.500.12657/39938>.

Scott, J.C. (1998) *Seeing Like a State: How Certain Schemes to Improve the Human Condition Have Failed*, Yale University Press, New Haven.

Stirling, A. (2010) 'Keep it complex', *Nature* 468: 1029–31 <https://doi.org/10.1038/4681029a>.

Vetter, S. (2005) 'Rangelands at equilibrium and non-equilibrium: recent developments in the debate', *Journal of Arid Environments* 62: 321–41 <https://doi.org/10.1016/j.jaridenv.2004.11.015>.

Vetter, S. (2020) 'With power comes responsibility – a rangelands perspective on forest landscape restoration', *Frontiers in Sustainable Food Systems* 4: 549483 <https://doi.org/10.3389/fsufs.2020.549483>.

CHAPTER 2

Decoding uncertainty in pastoral contexts through visual methods

Shibaji Bose and Roopa Gogineni[1]

Introduction

Pastoralists in popular culture have long been flattened, rendered in word and image as primitive, stuck in time; at once romanticized and vilified. In reality, pastoralist systems are highly adaptable and defy easy categorization, varying widely across and within regions. One commonality, however, is the presence of uncertainty. Pastoralists around the world constantly face and respond to the unknown and unexpected, from floods and droughts to locust plagues and market collapse (Chapter 1). Our work across six countries employed a variety of visual methods to surface and convey this diversity of experience.[2]

Since its invention in 1839, the camera has most often served as a tool for the elite (Berger, 1980). Its early use in police investigations, war reporting, and anthropological records was predicated on the belief that photography carries an incontrovertible truth. A photograph is a 'trace, something directly stencilled off the real, like a footprint' (Sontag, 1977: 154).

This assumption of veracity and the treatment of cameras as objective tools of documentation contributed to the widespread (mis)use of photography by researchers without explicit consideration of the practices by which images are produced and interpreted. In recent decades, however, the social and power dynamics inherent to image creation and consumption have been better explored, resulting in the development of more nuanced and critical visual methods (Spiegel, 2020).

The visual practices used in the PASTRES programme aim to capture the lived experiences of pastoralists. Through photos, videos, and narratives, the stories are largely told from the vantage point of pastoralists who are confronting diverse uncertainties. Storytelling through visual methods facilitates an engaged process of building knowledge that can eventually foster positive social change from below. In our work, it enabled pastoral communities to contribute experiential knowledge, providing an embedded understanding of uncertainty.

As a part of a broader agenda of community participation, stories shared through visual methods can help build critical consciousness to construct and forge knowledge and take action (Freire, 1970). Forging knowledge of place and facilitating engagement with local traditions and cultures can open a democratic space for dialogue among various climate actors. This provides an opportunity to support inclusion of local knowledge in policy, to systematize experience, and to draw out priorities for future actions.

Research design and methodological reflections

Visual tools for surfacing tacit and subjugated knowledge are increasingly used in a wide range of research activities, including in pastoral settings (Johnson et al., 2019). During the interaction, visuals provide a 'bridge' (Meo, 2010) that enables participants to converse about milieux that are very different from the researchers' own. The use of visual methods therefore helps the interaction between the researchers and the study participants. At a cognitive level, Pain (2012: 309) argues, 'because visuals use different parts of the brain than language, the two in combination could provide additional cues for understanding and encourage new connections between the two patterns of thought, thus facilitating new insights.'

This chapter reflects on a variety of visual methods selected to unpack multidimensional and evolving themes across a range of pastoral sites and therefore tries to make sense of the mangled, messy issues in different pastoral landscapes.

Photovoice

Letting go of 'researcher control' is itself an emancipatory approach. Visual tools collectively serve to centre the voices of pastoralists, inviting them to share beliefs and perceptions within their own frameworks of understanding and experiences of contending with unfolding uncertainties.[3] This embracing of the 'indigenous lens' helps orient the researcher to the intimate understanding of how pastoralists contend with variability, whether in terms of climate and weather or in relation to changing governance and market regimes. Across the case studies, what emerged was new hybrid knowledge and understanding of uncertainty refracted through perspectives of caste, identity, race, gender, and age.

In Amdo Tibet in China, for example, groups were formed in both research sites – Kokonor and Golok – and included faith leaders, women, and men from the Tibetan pastoralist community. In Isiolo, Kenya, one of the groups was composed solely of women of different ages, drawn from a mixture of social and economic circumstances to provide a distinct gender lens, while another was composed of younger men, offering an age-related perspective. In all sites, a diversity of views on how uncertainty as a concept was understood and how people responded to it was gleaned.

Visual methods help systematize local experience by appreciating participants' understandings of the local context and its socio-political and cultural elements. In Kenya, the methods led to a sense of empowerment among the participants, especially within the women's photovoice group. The participatory visual methods also help to lessen the power imbalance between the researcher and the photovoice group members, due to the collaborative nature of the method, which leads to the co-creation of evidence.

Due to the Covid-19 pandemic, digital media such as WhatsApp, Snapchat, and WeChat were used for remote facilitation of community-led methods, like photovoice, under lockdown. In Gujarat, India, this proved to be an effective method for a continuous conversation with the camel- and sheep-herders who travelled huge distances over the course of several months. In Amdo Tibet, China, remote conversation on these platforms helped the researcher to stay in contact with pastoralists (in both summer and winter pastures) in remote areas, with the exchange and dialogue continuing even during lockdown periods.

Social media ethnography

Pandemic-related travel restrictions disrupted the fieldwork of most researchers. Unable to carry out the photovoice exercises as initially conceived, the researcher in Tunisia observed images that proliferated in various pastoralist Facebook groups (Figure 2.1). This remote ethnographic approach shed light on diaspora networks and their sense of belonging. Across the sites, image

Figure 2.1 A collection of Facebook groups dedicated to Douiret in southern Tunisia. Members post archival photos, maps, poetry, obituaries, and live videos of sheep-shearing and olive-picking.
Credit: Linda Pappagallo.

Figure 2.2 Pastoralists fill jerry cans at the water pan in 1975 and 2020.
Credit: Gudrun Dahl (left), Goracha (right).

sharing on social media platforms offered insight into self-representation and reflected implicit ways of seeing.

Rephotography

Visual materials from archival sources allowed for interpretations of change across time by comparing images from today with those in the past. This proved a unique way of understanding the changes in pastoralists' space, identity, and sense of place, along with the evolving dynamics of uncertainty.

In Kenya, for example, photographs of a water pan taken 45 years apart were compared (Figure 2.2). The observed deterioration of the water pan prompted a conversation around the lack of community solidarity and its implications. When elders engaged in the discussion, they reflected on the need to mobilize a response to protect the shared resource.

Photo elicitation

Pictures taken by the pastoralists and/or by the researcher were able to draw out reflections around themes that were not easily expressed. This process uncovered subconscious and tacit knowledge of the unreliability and unpredictability of their setting.

The articulation of emotions became important, allowing people to reflect on how perspectives had been subjugated for social and economic reasons. This in turn provided a richer understanding of the evolution of their livelihood, food security, shelter, displacement, and identity, relating these changes to both personal and community experiences.

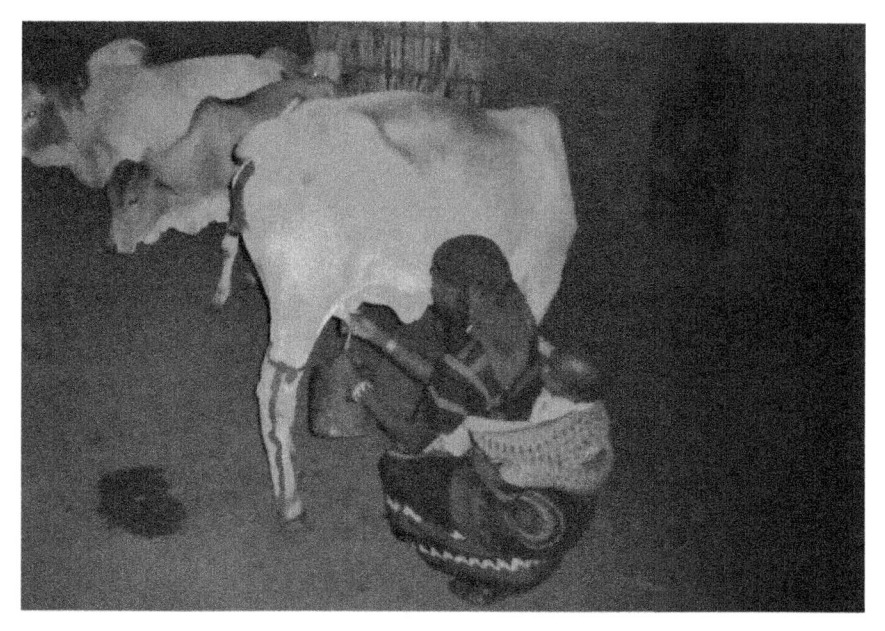

Figure 2.3 A wealthy adult male pastoralist took this photograph to show a 'natural dinner' for children. Others perceived it differently.
Credit: Malicha.

The lack of uniformity in the reflections of individuals and groups, even within a given context, mirrored the differences in their contextual experiences, depending on the vantage points of their lived daily experience. One of the most telling examples is a photograph of a woman milking a cow with a baby on her back in the early evening hours (Figure 2.3). To a middle-aged, relatively rich man, the photo symbolized the opportunity for nutritious milk from livestock, while the same photo shown to a poor, young female elicited empathy for the women burdened by her daily chores without a helping hand, a situation that may adversely affect her time for childcare.

In uncertain contexts, photo elicitation thus helped to draw out and prompt an in-depth line of thinking and contemplation.

Photo elicitation allowed the pastoralists in Sardinia, Italy, to express themselves through images from their personal archives. For example, a sheep-herder shared an old photograph of a meal shared with fellow pastoralists after a sheep-shearing session (Figure 2.4). This evoked in him memories of solidarity and continuity with the core traditions of the community, which he identified as a strong foundation against the uncertainties brought in by the market and Covid-19 (see Chapter 5). The discussions in Sardinia – as in the other sites – brought out pastoralists' engagements with dealing with uncertainty, surfacing memories, meanings, and often deep emotions about their landscape and their linkages with a fickle market.

Figure 2.4 An old photograph of a meal shared after sheep-shearing in Sardinia, Italy. *Credit*: Giulia Simula.

The community validation of the findings from photovoice groups in Amdo Tibet, China similarly provided a moment for deeper, collective reflection on pastoralism, uncertainty, and development. The discussions brought meaning to the evidence when participants shared photos and their narratives with the researcher. These conversations added trustworthiness and rigour to the research analysis.

Documentary photo/video by researchers and other interlocutors

Embedded photography and videography carried out by researchers, community groups, and occasional professional documentary photographers chronicled pastoralist communities in the face of evolving uncertainties, showcasing special events and everyday lives.

Often these pictures and films showed efforts to preserve significant customs and cultural moorings. Photographs from Sardinia, Italy, showing festivals that were attended by pastoralists dressed in traditional attire, highlighted the social value of maintaining cultural identities as pastoralists in the face of rapid change. In the same way, a video documentary of a yak beauty competition in Amdo Tibet, China, highlighted the importance of social processes at the heart of pastoralism, as well as the aesthetic and cultural features of livestock-rearing. Pictures from Gujarat, India, taken by a professional photographer working alongside the researcher, provide a fresh positioning, offering representations of how the camel and sheep-herders grapple with daily uncertainties during their opportunistic journeys to different lands.[4]

In Tunisia, the researcher worked with a local filmmaker to visit pastoralists and film interviews. Reflecting on the role of the camera, she noted that while its presence contributed to a degree of performance, the interviews were largely unstructured, often resulting in an uninterrupted monologue meandering into topics and avenues that may not otherwise have come up.

Visualizing uncertainty

In each of the study sites, the visual material complemented other research tools such as interviews or surveys and helped to open up discussions, providing an opportunity to articulate local conceptualizations of uncertainty. Conversations around uncertainty can often veer into murky abstraction, but photographs helped to ground the discussions in lived experience. The following sections offer some brief examples of how photographs encouraged debate about understandings across the sites.

Borana, southern Ethiopia

When asked about the locust plague that arrived in Borana in 2020, one pastoralist responded *'Bofa dheedhiitti, buutii afaan buune'* ('When we attempted to flee from a *bofa* snake (less deadly), we were met with *butti* (the most poisonous snake)'). Photos of locust swarms (Figure 2.5) provided a focus for discussion around the meaning of uncertainty. In the Borana pastoralist communities of southern Ethiopia, pastoralists explained that many terms

Figure 2.5 Locust swarms in southern Ethiopia.
Credit: Masresha Taye.

are used to express uncertainty, including *hinbanne* (not known or unknown), *haala* or *jilbii hinbanne* (limited knowledge, not knowing the likelihood), and *mamii* (unexpected).

It is believed that rain follows locusts. However, in 2020, rains preceded the arrival of a small locust outbreak. This provided an ideal environment for locusts to multiply and, later, they wreaked havoc on newly planted crops and fresh grassland. Pastoralists explained that they reoriented their strategy to deal with the situation, but mobility – a key feature of the pastoral coping strategy – was halted to contain the spread of Covid-19. This combination of shocks devastated many pastoralists in Borana (see Chapter 7).

Amdo Tibet, China

Kokonor is a sacred lake on the Qinghai plateau in Amdo Tibet. In 2016, the lake began expanding, subsuming the winter pasture of many pastoralists. As Uncle Lhabe explained when describing the photo shown in Figure 2.6, 'That is my winter house, we spend most of our time there. Unfortunately, I lost my winter pasture and the house in 2019 due to the lake expansion. Now I need to rent pasture from other pastoralists, and I need to find a place to stay.'

Discussion of the lake expansion provoked many discussions of uncertainty. As noted in Chapter 4, the Tibetan phrase *Bsam yul las das pa* means something beyond one's imagination, beyond the realm of thought. It then

Figure 2.6 Uncle Lhabe looks out at his former winter pasture, now underneath the lake.
Credit: Palden Tsering.

is unmeasurable, unpredictable, unexpected, and impossible to prepare for. The lake expansion was definitely in this category of experience. As another pastoralist explained, 'What is going to happen is unpredictable; all we can depend on is the present, what happens now, we deal with it now.'

During a photovoice feedback session in Kokonor, various interpretations emerged regarding the causes of the lake expansion. Some described the occurrence as the intended result of successful conservation efforts carried out by the government. Others pointed to religious reasons, saying it was the consequence of neglecting certain religious practices, such as releasing fish or failing to carry out the water-deity ceremony. Participants also identified climate change – marked by increased temperatures, melting glaciers, and changing precipitation levels – as a contributing force. Discussion of the photos collected by the men's and women's photovoice groups allowed the community to reflect on the different causes and consequences of the lake expansion, and the impacts on pastoralists' lives of the uncertainties.

Isiolo, northern Kenya

Ali Saleisa, a resident from Merti in Isiolo County, northern Kenya, shared images of his dead goats, which he had photographed to seek compensation for losses (Figure 2.7). The photographs embody intersecting uncertainties faced by pastoralists in the region: wildlife attacks, the impact of

Figure 2.7 Goats lost to a lion.
Credit: Ali Saleisa.

invasive species, conservation policy, and the role of the local government, he explained.

> When livestock disappear, they are not easy to recover; the entire area is covered by the thorny *mathenge* (*Prosopis juliflora*) and other shrubs. We took the flock to a place called Goo'aa near the *galaan* (river) to feed on a *d'igajii* shrub, which is not as thorny. The goats dispersed and a lion attacked and killed 84 goats. We kept searching with the help of other herders but only found the carcasses; the lion killed them all without eating any flesh! It was very disheartening. I reported this to the chief, received a letter, and followed it up with the county government. It has been five years since then, and I have never been compensated.

Discussions such as this, prompted by the sharing of some images, raise many questions about how uncertainties are understood, experienced, and confronted socially and practically (see Chapter 6). Further discussions with elders highlighted local conceptions of uncertainty: 'No human knows everything, even if you are an expert in one area', explained one elder in the Waso Borana community. As Abdullahi Dima observed, '*Hooraa buulaan* (whoever desires a prosperous life) will survive all the *c'inna* (calamities) that we face in society.' Individuals who desire a successful life are always in fear of the unknown ('*Hooraa buulaan waa soodaataa*') and must know how to handle those fears without harming others, he explained.

Kachchh, Gujarat, India

When asked if they could predict certain events, like the pandemic or the first rainfall, or whether they knew to expect such events, Rabari pastoralists in eastern Gujarat shrug their shoulders and say '*Kone khabar*' ('Who knows?' 'We don't know', and 'Only God knows'). Rather than speculate on events that are beyond their control, the Rabari often say '*Thaay tyaare hachu*' or 'It will be true when it happens'/'It will be known when it happens'.

As discussed in Chapter 3, Rabari pastoralists have long responded to unfolding circumstances through movement. Their mobile camps are packed on camels, tractors, and tempos (small trucks or vans), moving daily and seasonally to exploit emerging opportunities. Figure 2.8 shows one photo that reflected discussions on pastoral mobility. As a young male Rabari pastoralist explained:

> We decide two to three days in advance of moving. We think of where there is good grazing, where it is worth staying. We may stay in one place one night or even 15–20 days. We contact pastoralists ahead and farmers ahead – they tell us, come to our farms in a week or come in 10 days. Our *mukhi* or leader goes to find grazing. This scouting is called

Figure 2.8 A Rabari camp on the move.
Credit: Natasha Maru.

niharu karvu. If we take a stroll through the outskirts of a village, then we know that there are crops that will be harvested in so many days and will be available for the livestock to graze. It is not decided where we go, but we have built relationships in certain villages over the years where we are comfortable, so we try to go there year after year, providing grazing is available.

Douiret, southern Tunisia

When asked what comes to mind with the word 'uncertain' (*al majhoul* or *aashakk*), one pastoralist in the mountainous village of Douiret in southern Tunisia responded as follows:

> When there is no rain it's hard. It's quite difficult. Thankfully, I manage the livestock on my own. If I ask a herder to be responsible for the herd, along with the increase in forage and barley prices ... I wouldn't make it ... I do not have the resources to resist. I contribute with my own money, my family is helping me a bit, we rise and fall, and we thank God. This is uncertainty. One day the livestock eat, the next day they don't. And hopefully, since we have in our pocket some money, we will provide the flock with what's necessary for their needs. Living is hard, especially when there is no rain, it's hard. Along with the high cost of living, in our time it got harder.

Figure 2.9 An extended family gathers during a sheep-shearing in Douiret, Tunisia.
Credit: Linda Pappagallo and Hamdi Dallali.

In Douiret, many pastoralist families navigate such uncertain circumstances by maintaining transnational networks. Douiris travel to Tunis or abroad, all while financially contributing to the keeping of herds back home (see Chapter 8). An estimated 80 percent of Douiris do not live permanently in the village but instead return seasonally, often during the *jezz*, or sheep-shearing, season. This marks the beginning and the end of the pastoral year in the southern regions of Tunisia. Extended family members gather to help shear the community's stock, as shown in Figure 2.9. These are moments of discussion, and one returning pastoralist reflected on the theme of migration and uncertainty:

> This coming and going creates a sort of continuity for me, for my mother and my younger sister. This continuity manifests itself every time someone comes, when we receive letters, when someone says that it has rained in Douiret. So, this relationship with the village is virtual, not real, and manifests itself with people that come, with couriers, with good and bad news. We continue to live the village in an imaginary fashion, in a virtual fashion, we continue to keep the images of our childhood vivid. The caves, school, the olives, the dromedaries.

As discussed earlier, the 'imaginary' or 'virtual' presence is maintained through active online communities. Dozens of Facebook groups and pages are dedicated to Douiret, with members posting photos and videos of sheep-shearing and olive-picking, images used to establish presence in spite of physical absence.

Sardinia, Italy

According to pastoralists from Sardinia, precarity and uncertainty are the only constants. As they explained, when we make plans, we rarely say 'Yes'; instead, we say, 'Barring unforeseen events, I will be there!' A pastoralist from the south of the island explained how uncertainties about price, demand, supply, and market conditions are endless: 'Uncertainty is there every day ... uncertainty around whether to continue or not because it is becoming less and less profitable and difficult to predict from an economic point of view.' Although 'uncertainty' in Italian is translated as *incertezza*, other terms are also used in the Sardinian context, including *precarieta* (precarity), *insicurezza* (insecurity), and *imprevisto* (unexpected event), reflecting the livelihood contexts of uncertainties in a pastoral setting.

Although pastoralism is central to the livelihoods of most Sardinians, the form it takes varies dramatically. Some make and sell cheese on their own, others increase bargaining power by selling milk through a cooperative. The constant fluctuation in markets requires close attention and adaptation. Reflecting on the photo in Figure 2.10, Felice explained how milk prices fluctuate a lot and the best solution for him is to make cheese and sell it, gaining greater income and keeping the farm going (see Chapter 5).

Figure 2.10 Felice's cheese production, Sardinia.
Credit: Giulia Simula.

The afterlives and circulation of visual material

Beyond a tool to surface knowledge and open discussions in pastoralist communities, the visual methods used across the sites also produced a collection of images that clearly communicated salient findings, visualizing uncertainty and making what is otherwise a quite abstract concept real. In order to share the insights from the six sites more widely, the photographs and their linked narratives were curated and shared through a variety of platforms, including travelling in-person exhibitions, an online exhibition, and photo newspapers.

Initially, researchers had intended to hold exhibitions in each of the field sites, but pandemic-related restrictions around travel and gathering disrupted some of these plans. In response, hand-held exhibitions were designed in the form of photo newspapers. Each tabloid-size paper contained a series of photographs that distilled central research themes in an accessible format. Although the content was primarily visual, accompanying text was translated and printed for local audiences. The research culminated in six site-specific newspapers and one longer thematic newspaper exploring manifestations and responses to uncertainty across sites. The newspapers were widely shared and received much interest from the pastoral communities.

Researchers employed the newspapers in a variety of ways. In Ethiopia, Borana language newspapers were brought to homesteads that participated in the research. In Amdo Tibet, newspapers were distributed at a horse festival in Kokonor (Figure 2.11) and used in classrooms. Newspapers were hung on clotheslines for an impromptu exhibition in Isiolo, Kenya, during a feedback

Figure 2.11 Tibetan language newspapers distributed at a horse festival.
Credit: Palden Tsering.

Figure 2.12 Swahili language newspapers on display at a community feedback session in Isiolo.
Credit: Ian Scoones.

meeting bringing together pastoralists and public officials (Figure 2.12). This process of sharing and receiving feedback through exhibitions and small group meetings demonstrated the importance of pastoralist knowledge in the research.

The inaugural cross-country exhibition titled Seeing Pastoralism took place outside Alghero in Sardinia, Italy in September 2021. Newspapers and prints were hung on yarn spun from Sardinian wool in the gardens of an *agriturismo* hotel (Figure 2.13). The event attracted hundreds of visitors, including many pastoralists from nearby communities as well as international guests. After the exhibition, one Sardinian pastoralist reflected:

> It has been very interesting to take this virtual walk through the different sites. The reflection I can make is that, notwithstanding the different environments, climates, and cultures, there is a single thread that links pastoralists across the world. And that is adaptation. I believe that this disposition of adaptation that pastoralists have is in fact a reflection of the adaptation that animals themselves have. In reality, animals adapt and we adapt along with them … these realities represent issues that do not only concern pastoralists because we are talking of climate change and global economy; these are aspects that concern everyone.

Figure 2.13 Sardinian pastoralists looking at photos of Tunisian pastoralists.
Credit: Roopa Gogineni.

Following the successful launch in Sardinia, the Seeing Pastoralism exhibition travelled to the international climate meeting, COP26, in Glasgow, the Global Land Forum in Jordan, the Stockholm+50 conference, the online European Development Days Forum, and events in Addis Ababa, Bhuj, Brighton, Brussels, Isiolo, Florence, and southern Tunisia. A digital exhibition was designed and published, incorporating audio and video content in addition to the photographs and text narratives.

The site-specific and cross-country exhibitions, both in-person and online, have generated a reflective and socially critical dialogue between community members, researchers, and policymakers about pastoralism, uncertainty, and development in ways that could not have been imagined had more conventional research formats been used.

Conclusion

This multi-country experience has demonstrated how a hybrid visual approach may be used to surface and share knowledge about complex and intersecting uncertainties. Much like the pastoralists at the centre of this work, the researchers themselves faced uncertainty, which sparked experiments and innovations in the use of visual tools.

To understand uncertainty from the eyes of the pastoralists has always been a challenge to the traditional researcher aiming to build research credibility, give back the results of the research to the communities at the margins, and build knowledge together. Participatory visual research methods were able to unearth hidden tensions in uncertain pastoral landscapes. These methodologies, which were accessible to diverse groups – including women and youth as well as older men – provided insights into their own world-view, going beyond 'literary' or 'reflexive' approaches. By shifting authority and power in the process of knowledge-making, the approach took inspiration from feminist, post-colonial, and critical epistemologies (Denzin and Lincoln, 2003). The differentiated knowledges and perceptions reflected in people's explorations of uncertainty reinforced the argument that there are multiple realities, which are socially constructed, and these must be engaged with in processes of development (Yilmaz, 2013).

While the legacy of lens-based tools in academic research is fraught, the potential of thoughtfully designed non-didactic visual methods is vast. Such tools may challenge the narratives of the dominant frames of development planners, policymakers, and implementers with the more tacit, hidden knowledge drawn from the daily rhythms of the lives and livelihoods of the pastoralists. In so doing, such approaches can help open up debates about pastoralism, uncertainty, and development in ways that are not constrained by conventional understandings.

Notes

1. This chapter was developed collaboratively with PASTRES researchers Natasha Maru, Tahira Mohamed, Linda Pappagallo, Giulia Simula, Masresha Taye, and Palden Tsering who are authors of the six subsequent chapters and worked across the six sites discussed.
2. Now curated into an online resource and exhibition, Seeing Pastoralism (seeingpastoralism.org).
3. This effort built on previous photovoice work with pastoralists; for example, in Kenya, Uganda, Sudan, and Ethiopia (https://fic.tufts.edu/wp-content/uploads/Pastoral-Visions.pdf) and Mongolia (https://documents1.worldbank.org/curated/en/986161468053662281/pdf/718440WP0P12770201208-01-120revised.pdf).
4. Photographer Nipun Prabhakar (nipunprabhakar.com) documented the migration of Rabari pastoralists to mainland Gujarat and facilitated photography workshops.

References

Berger, J. (1980) 'Uses of photography', in J. Berger, *About Looking*, pp. 48–63, Pantheon Books, New York NY.

Denzin, N.K. and Lincoln, Y.S. (2003) *The Landscape of Qualitative Research: Theories and Issues*, 2nd edition, Sage, Thousand Oaks CA.

Freire, P. (1970) *Pedagogy of the Oppressed*, Herder and Herder, New York NY.

Johnson, T.L., Fletcher, S.R., Baker, W. and Charles, R.L. (2019) 'How and why we need to capture tacit knowledge in manufacturing: case studies of visual inspection', *Applied Ergonomics* 47: 1–9 <https://doi.org/10.1016/j.apergo.2018.07.016>.

Meo, A.I. (2010) 'Picturing students' habitus: the advantages and limitations of photo-elicitation interviewing in a qualitative study in the city of Buenos Aires', *International Journal of Qualitative Methods* 9: 149–71 <https://doi.org/10.1177/160940691000900203>.

Pain, H. (2012) 'A literature review to evaluate the choice and use of visual methods', *International Journal of Qualitative Methods* 11: 303–19 <https://doi.org/10.1177/160940691201100401>.

Sontag, S. (1977) *On Photography*, Farrar, Straus and Giroux, New York NY.

Spiegel, S.J. (2020) 'Visual storytelling and socioenvironmental change: images, photographic encounters, and knowledge construction in resource frontiers', *Annals of the American Association of Geographers* 110: 120–44 <https://doi.org/10.1080/24694452.2019.1613953>.

Yilmaz, K. (2013) 'Comparison of quantitative and qualitative research traditions: epistemological, theoretical, and methodological differences', *European Journal of Education* 482: 311–25 <https://doi.org/10.1111/ejed.12014>.

CHAPTER 3
Engaging with uncertainties in the now: pastoralists' experiences of mobility in western India

Natasha Maru

Introduction

I first participated in a migration journey in November 2019. I was travelling with Vasu's flock from one village to the next, just a short trip. Vasu belongs to the Rabari community of mobile pastoralists from Gujarat in western India. She, along with her brothers Varna and Veho, travels with their flock of sheep throughout the year to ensure fresh and nutritious fodder for their animals. They spend the monsoon months in the commons of Kachchh District that sees fresh grasses after the rains, while grazing on crop residues in agricultural hotspots in central Gujarat in the summer and winter months once the grass dries out.

We started our journey from Vasu's home village in Kachchh. Having grazed in the grasslands of northwest Kachchh, they were now making their way towards central Gujarat. After walking for about an hour, we arrived at a suitable site in the bush near the next village, and the women began setting up the camp. The Rabari sleep under the stars with only a few belongings transported on top of the camels. The camp consisted of a couple of charpoy beds, cooking utensils, and their minimal belongings – such as clothes, toiletries, tools, and medicines – all neatly organized on and around the beds.

But just as soon as the camp was set up, we found ourselves caught in a violent and unexpected storm. Huge hailstones – larger than I had ever seen – pelted the ground and rain continued well into the night. The storm caught us completely by surprise! While normally the men would sleep on

Figure 3.1 The morning after the hailstorm.
Credit: Natasha Maru.

the ground close to the sheep pen, this night we all huddled on the damp charpoy beds under a flimsy tarpaulin roof for shelter. The night was wild. As the storm blew outside the tent, a wild dog took away a lamb from inside the tent despite our vigilance.

It was well past the monsoon season in the region and this late and violent rain destroyed the *Bt* cotton sown for a winter harvest. This would mean a huge loss for the farmers as they would miss out on a season's earnings, having already invested in seeds and inputs for the crop. Therefore, the state government announced a subsidy on inputs for the farmers and encouraged them to resow the crop. The availability of irrigation through infrastructural development also meant that resowing out of season was possible, and the farmers were able to have a new crop to minimize their losses.

The resowing shifted agricultural timelines, extending the impact of the rains into the next season. As pastoralists do not graze on standing crops, Vasu's camp was forced to slow down and circle around the fallow fields in Kachchh for several months after the storm in anticipation of the harvest. They were still in Kachchh in February 2020, frustrated as they waited for the crops to be harvested so they could graze their flocks and travel on.

Uncertainty and change lie around every corner for the pastoralists, whether through unexpected extreme weather events (like hailstorms), the dogs who come and take away lambs in the night, or the organization of a life on the move. Uncertainties exist at various scales. For example, the dog attack happened in a single night at the spatial scale of the camp, while uncertainties related to harvests and access to crop residues for livestock fodder are regional and seasonal in scale. What is evident from the experiences of Vasu's family is that just as change is the only constant, uncertainty is the only surety.

Pastoralists in western India have a long history of managing change and uncertainty through their mobility. Whether it is variable environments, disease outbreaks, or fluctuations in market conditions, pastoral systems have evolved strategies, processes, and institutions for mobility to engage with uncertainties. They are increasingly being valorized for their capacity to adapt to, take advantage of, and overcome difficult circumstances (FAO, 2022). This view counters the dominant, and persistent, narratives of vulnerability, struggle, and coping with 'harsh' and 'scarce' environments, which consequently underpin mainstream imaginaries of pastoralists as 'backward', unproductive, and marginal (Butt, 2016).

Yet such views of successful adaptation have been challenged for being apolitical, ignoring the increasing pressures that pastoralists face, as shown by the frustration felt by Vasu's family towards the end of the vignette. Whether the challenges relate to anthropogenic climate change or a shift in political economy geared towards more commercial crop-farming and industry, pastoralists in western India, as elsewhere, bear a disproportionate burden of adaptation.

In this chapter, I explore these tensions through a temporal lens to introduce a new way of looking at mobility and uncertainty, and the limits of pastoralists' practices, based on a case study of the Rabari from Kachchh District in Gujarat, India. Uncertainty can be understood in two ways: empirically, in the sense of uncertain events and circumstances, and as a strategy applied by pastoralists to adapt to new circumstances. I see variability and change as intrinsically temporal, and these temporalities as central to pastoralists' mobile practices, social relations, and institutions.

Highly attuned to shifts in their context, the pastoralists display flexibility and agility to adapt to changes in the here-and-now. The 'pastoral present' thus serves as the arena for action where pastoralists engage with uncertainties as they unfold; that is, in the now. Rather than a discrete period, the 'now' draws on experiences from the past to address expectations of the future. Juxtaposed against these synchronized temporalities of pastoral mobility is the rapidly shifting political economy of Kachchh modelled on fast-paced neoliberal capitalism and globalization. This limits pastoralists' abilities to respond adaptively to uncertainties, with adverse policies undermining pastoralists' strategies.

While, on the one hand, pastoralists are being lauded for their ability to counter mainstream narratives of unproductivity and destitution, on the other hand, their capacities are challenged as shifts in political economy fail to account for pastoral livelihoods.

The Rabari: their context and pastoral practices

The Rabari describe their origin in a religious myth where the Hindu god, Lord Shiva, created their forefather to take care of his wife's camels (Westphal-Hellbusch and Westphal, 1974). They are a prominent Hindu community from the western Indian states of Gujarat and Rajasthan, with some small groups settled in central India. Many of the Rabari now keep small ruminants instead and remain attached to their original identity as *maldhari* or pastoralist.

Rabari pastoralists are specialized livestock producers who are adept at taking advantage of variabilities in the distribution of fodder resources in order to secure optimum nutrition and wellbeing for their flock (FAO, 2021). They come from multi-ethnic, multireligious, spatially segregated villages and have no territory of their own (Mehta, 2005). Therefore, they rely on mutually beneficial relationships with other groups such as farmers and local livestock-keepers to graze on grassy commons and crop residues after harvests. They mostly graze on cotton crop residue in the winter months and wheat crop residues in the summer months, along with those of other crops such as castor, sesame, *moong* beans, pigeon pea, sorghum, and cumin (Figure 3.2).

The pastoralists journey across the landscape in migrating groups of two to ten flocks. Timing is very important for Rabari pastoral practices and their mobility. The Rabari time their movements paying attention to several rhythms, including crop lifecycles from sowing to harvest, animal lifecycles, and lambing periods. Weather, especially rainfall, has a critical role in these cycles. The migrating group, as a social institution, has imbibed these temporalities. It remains agile and can easily change labour and livestock composition to adapt to the available fodder resources. These temporal aspects of mobility are thus key to how pastoralists respond to uncertainties.

In recent decades, Gujarat has undergone some major shifts in its political economy. Following a massive earthquake in Kachchh in 2001, the region has seen statist developmentalism where major investment has come to represent progress and modernity. Huge areas of the sparsely populated region have been opened for large-scale export-oriented industries and commercial agriculture, making Kachchh one of the fastest growing regions of India in economic terms. Such developmentalism seeks to instil stability, order, and productivity within a landscape that has long been considered wild, empty, and 'waste' (Bharwada and Mahajan, 2006).

Most relevant for the pastoralists are the developments in the agricultural sector. Genetically modified *Bt* cotton was introduced in India in 2002 and soon became one of the biggest crops in Gujarat. Aided by new irrigation possibilities, including through India's largest dam project, the Sardar Sarovar Dam, along with the subsidized supply of synthetic fertilizers, this has changed the region's agrarian landscape in profound ways. These developments have had major implications for the temporalities of agrarian life: accelerated harvests have been achieved by manipulating crop lifecycles, through seed hybridization, and weather cycles have been changed through dam construction.

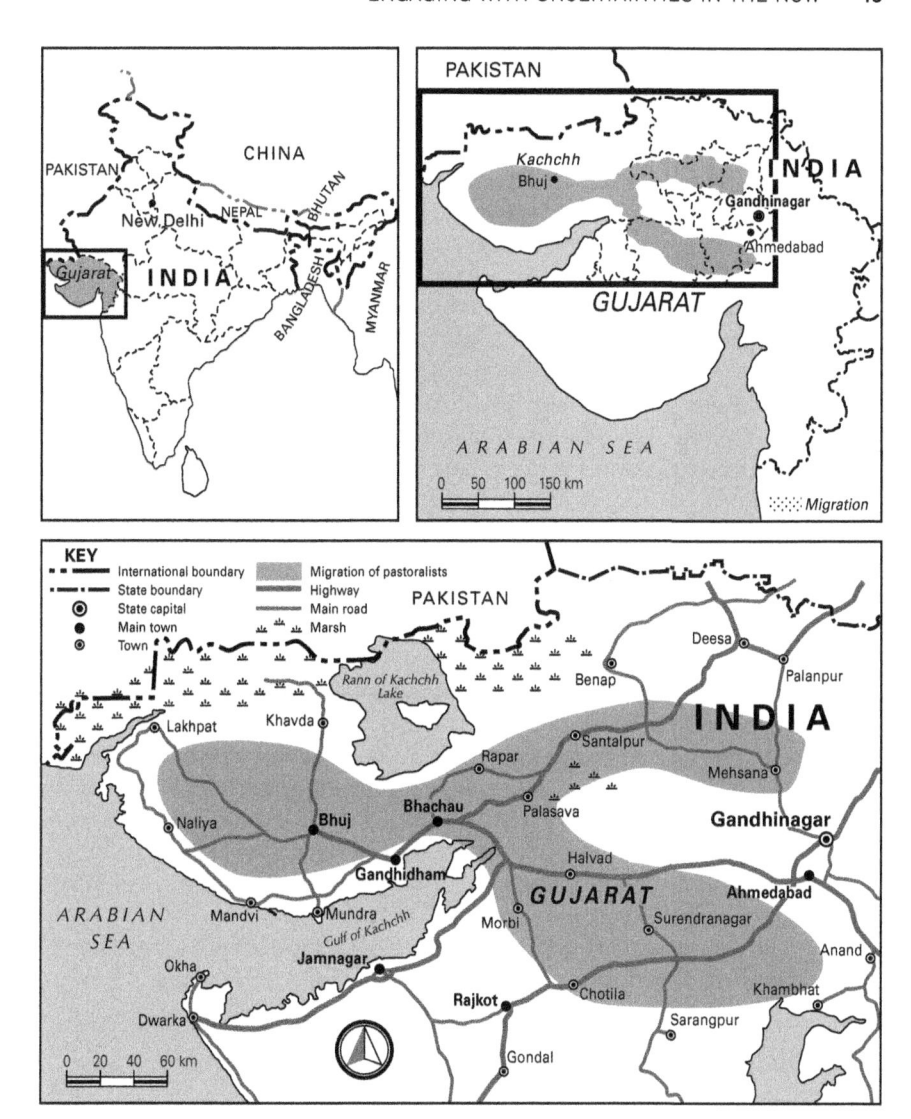

Figure 3.2 Map of the Kachchh study area

Although these changes seem to favour industrialization and settled agriculture to the detriment of pastoralism, pastoralists have found ways to take advantage of such capital-driven developments. For example, through the strategic pacing of their mobility in interaction with agrarian dynamics, Rabari pastoralists have been able to avoid travelling long distances. They receive not only nutritious protein-rich fodder, in the form of cotton crop residues, but also income for the manure from their animals from farmers who increasingly have the potential to pay and see value in organic fertilizers.

At the same time, the pastoralists have become valuable contributors to global livestock chains by meeting growing demands for animal source foods. Nearly 700,000 live sheep and goats are exported from India each year, mostly to the Middle East, making it a lucrative profession (TNN, 2020). In fact, the Rabari pastoralists are even adapting their animal breeds in response to these shifts as the wool market has declined in favour of the meat market.

Therefore, the pastoral production system remains tied to capitalist development and has adapted to changes over time in a way similar to the Rabari of Rajasthan as described by Robbins (2004). For example, while agriculture policies in western India neglect pastoralism and promote commercial crops, pastoralists adapt to these changes by synchronizing their movements with harvest times. They not only receive fodder access but also capitalize on the growing agrarian market by earning income through manure exchange. While changes in political economy seek to push pastoralism outwards to its margins, it grows inwards towards the centre of the regional economy. Rather than subvert or submit to changes emerging from capitalist development and globalization, the pastoralists engage with these processes dialectically, 'not only to enhance the possibility of survival but also to reclaim the material and symbolic conditions for the flourishing of life' (Kolinjivadi et al., 2020: 907).

Yet, although the pastoralists remain economically relevant, they continue to face social marginalization. This has many implications. Locally, religious proscription in the region affects them economically by interrupting the sale of animals for slaughter and also asks them to justify socially their *paap no dandho* (sacrilegious work). In a broader sense, pastoralism continues to be seen as out of kilter with the dominant paradigm of progress, modernity, and development, leading to the neglect or active marginalization of pastoralists. Given these changing contexts, there are limits as to how far the pastoralists can be expected to adapt. Understanding this requires a deeper unravelling of the dynamics of uncertainty and its implications for development.

Uncertainty, mobility, and the now

'Tamne khabar hati aavu thashe?' ('Did you know this would happen?'), I asked Harjanbhai, the elderly leader of Vasu's migrating group, about the hailstorm. We were sat under a tarp, tied loosely over the charpoys for shelter, along with a few other members of the group. Harjanbhai, as the leader or *mukhi* of the group, was responsible for securing grazing for the five flocks that had chosen to accompany him: *'Kon jaane?'* ('Who knows?'), *'Amane kem khabar hoy?'* ('How would we know?'), *'Bhagwan jaane'* ('God knows'), *'Kai nakki na kehavaay'* ('Nothing can be said for certain').

In fact, these are the kind of responses I received in many different circumstances. For example, the pastoralists change group configuration throughout the year in response to contextual specificities. But when asked who they were planning to move with in the next season, they would indicate ambiguity by saying '*E toh kai nakki na kehavay*' ('That cannot be said for certain.') They might, in practice, travel with the same group year after year but they keep their options open to be able to adapt to variable conditions as they encounter them. Embracing uncertainty and allowing for multiple possibilities, as reflected in discourse, is key to maintaining the flexibility integrated within the social institution of the migrating group, and this allows them to adapt to uncertain events through their mobility.

Similarly, Vasu knows that travelling to Gujarat with Harjanbhai is a possibility, but that choice is not made until closer to the time of departing from Kachchh, once fodder conditions are known. Therefore, there is always a temporal horizon to uncertainty. By this, I mean the temporal distance within which objects and events come to influence behaviour such as mobility. This is not a discrete period along a linear timeline of past, present, and future but, rather, a temporal consciousness within which experiences of the past, and aspirations, expectations, and fears for the future intertwine with and shape the present. Rather than remaining separated from the past and future, the pastoralists act in an ever-evolving now. The temporal horizon of the 'nomadic present' or now means that reasoning moves between knowledge from past experiences and future expectations by 'tacking back and forth between nitty-gritty specificities of available empirical information and more abstract ways of thinking about them' (Adams et al., 2009: 255). It is therefore a 'constitutive and productive heterogeneity, a circulation of multiple times in a single instant' (Luckhurst and Marks, 1999 quoted in Burges and Elias, 2016: 4). Therefore, the now is the contingent cumulation of history and context rather than an abstracted and discrete moment.

The Rabari pastoralists have a long history of managing change and variability. Their temporal horizon remains highly attuned to shifts in context. They do not seem to get bogged down in speculation about things they are unable to control or foresee. Rather, they rely on past experience and their networks to assimilate knowledge on the go, in the here-and-now, as events unfold. They operationalize this knowledge in the form of mobility decisions that ensure the best outcomes for their flock. For example, given the available information regarding weather and fodder conditions in the area, as well as the labour capacities across the migrating group, Harjanbhai decided that it was time for his migration group to set off on their journey. They were, firstly, able to manage the composition of their group and, secondly, decide on where and when to go, having scouted the area and asked for information from a range of contacts.

At the same time, the group confronted several uncertainties. While they were aware of the occurrence of hail as a weather phenomenon, it was completely unexpected that day, and its likelihood could not have been predicted. Similarly, while they were aware of the possibility of a wild dog taking away their lambs, the likelihood of this happening was completely skewed by the weather event – instead of keeping the lambs in the pen with the adult animals, the lambs were kept under the tarp to save them from getting wet and ill. The menfolk were also sleeping under the tarpaulin, instead of keeping guard close to the pen, and the storm obscured the vision of those who might have otherwise seen a roving dog.

Following the events described, the group decided to move to a different area the next day – somewhere stonier so that wet soil would not hamper the sheep's movements. This demonstrates the pastoralists' ability to read the environment and understand the interaction between natural elements such as soil and rain. They relied on their experiential knowledge of the terrain to decide on their moves. It reflects their contextually embedded flexibility, the agility, the 'built-in elasticity' to perceive and respond immediately to change (Hazan and Hertzog, 2011: 1). Such real-time responsiveness relies upon the mobility of the system, characterized by mobile assets – the animals, the leanness of the camps, knowledge of the terrain, and the social relations built over generations.

The role of the *mukhi* or the leader is crucial in these decisions. *Mukhis* such as Harjanbhai are self-appointed leaders of a migrating group who take charge of ensuring fodder availability for a group of flocks under their charge. The *mukhi* is familiar with the landscape and has the social resources and communication skills needed to negotiate and make decisions regarding fodder access. He understands not only the 'right' place to move to but also the 'right' time to move. He has a kairological understanding of time based on past experiences and intuition. In contrast to the abstracted, objective, and impersonal time of chronology or *chronos*, *kairos* is the perception of 'temporal opportunities' (Maffesoli, 1998: 110), 'the moment that must be seized' (Ingold, 2000: 335) and 'time considered in relation to personal action, in reference to ends to be achieved in it' (Robinson, 1950 quoted in Jacques, 1982/1990). The *mukhi* grasps opportune moments and makes decisions that are both calculated and spontaneous, informed as well as instinctive.

As such, the *mukhi* thus performs the role of a 'reliability professional' (Roe, 2020) who manages uncertainties by employing various mobility strategies to reduce variability. He is able to assimilate knowledge from multiple sources and scales across networks, bringing core considerations within the temporal horizon. He has a step firmly in the present with a view towards the future, while relying on past knowledge. Moreover, he acts as a broker, mediator, and negotiator within networks, relying on experience, tacit knowledge, diverse senses, emotional intelligence, intuition, and networking.

The migrating group also integrates temporalities within its social institution as it shape-shifts to reflect mobility decisions. While Harjanbhai's group was experiencing the hail away from their home village, another migrating group, that of Nathubhai, stood at the outskirts of their village. Nathubhai's camel drowned and died unexpectedly while grazing in the mangroves close to the village. Unable to travel long distances without the help of a camel to carry their belongings, Nathubhai was forced to remain around his village where he could rely on the support of friends and relatives to move camp. This also meant that he chose to remain with a single flock rather than join others in a group. A single flock is able to sustain itself within a limited area as its fodder requirements are lower than a large group. The modularity and dexterity of the migrating group supported Nathubhai in facing the uncertainties emerging from the death of his camel and to modify the spatio-temporal organization of his mobility to meet fodder availability.

The same property of a migrating group was operationalized by Vasu. Her father had unexpectedly passed away in the bush the previous year, leaving her and her brothers responsible for taking care of their animals. While they normally travelled with their uncle Hemangbhai's migrating group, this year, they had chosen to go with Harjanbhai instead, changing the composition of both groups. While such choices call for delicate social navigation, the institution of the migration group allows members to freely exit and enter a group at any time and is adaptable to such changes, also indicated through the phrase *kai nakki na kehavay* (nothing can be said for certain).

Thus, mobility and its temporalities are key to pastoral adaptation to uncertainty in western India at various spatial and temporal scales. The practices, social relations, and institutions of mobility are flexible, prompt, and modular in design to enable the pastoralists to adapt to new and unknown circumstances as they emerge. Being so attuned means that rather than following a linear path, the pastoralists embrace uncertainty as a strategy and act in response to an ever-changing present.

Limits to adaptations to uncertainty

The previous section has highlighted the adaptive capacity of Rabari pastoralists to respond to uncertainties across space and time through movement. Crucial to this are real-time adjustments made in the now. A focus on the nomadic present therefore enhances the agility of pastoral systems and opens alternative possibilities.

Understandings of pastoralism through an uncertainty lens challenge mainstream framings that see pastoralists as 'outdated, irrational, stagnant, unproductive, and ecologically damaging' (Butt, 2016: 463). Such a perspective shows the capacity of pastoralists to not only manage but also take advantage of variable conditions (Krätli and Schareika, 2010). Yet while, on the one hand, pastoralists are being lauded for their flexibility and mobility, on the other, their capacity to adapt is challenged further as shifts in political economy fail to account for pastoral livelihoods. Despite growing recognition of pastoralism within international development as both economically viable and environmentally beneficial, the 'sediment of nomadism' (Kaufmann, 2009) continues to undermine pastoralism and privilege linear visions of modernity, development, and progress.

In Kachchh, such developmentalism has led to the structural oppression and marginalization of pastoralists through adverse policies, as discussed earlier. The transformation of the political economy in Kachchh has rendered the pastoralists out of kilter with, and irrelevant in, popular imaginaries of progress and development. The temporal horizon within which pastoral action is oriented is increasingly being disrupted through shifts in political economy. Borrowing from Rosa (2013), the Rabari pastoralists can be seen as facing three levels of acceleration – technological acceleration, which is evident in transportation, communication, and production; acceleration in social change, in cultural knowledge, social institutions, and personal relationships; and acceleration in the pace of life – that are transforming the temporal structure of society. Such accelerations are shifting the template of social life, such that the past can no longer inform the future, leading to a 'contraction of the present' (Rosa, 2013: 76).

Bauman (2000), too, suggests that the fluidity of a 'liquid modernity' is disrupting the templates that govern social life. In this *zadapi zamano* (fast age) 'only the sky (or, as it transpired later, the speed of light) was now the limit, and modernity was one continuous, unstoppable and fast accelerating effort to reach it' (Bauman, 2000: 9). Similarly, Guyer (2007) shows that there has been an evacuation of the near past and the near future in favour of significant short- or long-sightedness as contemporary modernities recast temporality.

These shifts have significant implications for Rabari pastoralism. Such changes in temporal horizons disrupt nomadic temporalities, limiting the pastoralists' capacity to adapt. For example, Nathubhai's eldest son, Valo, was never sent to school and had been sheep-herding since he was young. Yet now, given the negative discourses surrounding pastoralism, he wishes to exit and try his hand at a new profession. With greater access to information and images of quick success floating in through his smartphone, his vision has expanded from the immediate shifts in his surroundings to longer-term concerns about the future.

He has tried to exit pastoralism and gain access to several alternatives, such as setting up a dairy or working as a taxi driver, or taking up a job in a factory, but his attempts have been unsuccessful. He has returned to the animals each time as pastoralism continues to be economically viable, even if politically marginalized. At the same time, while he faces precarity regarding his job and future in the short term, his lack of education being a long-term disadvantage, his migrating group suffers as well. In the short run, they are unable to operationalize strategies like herd-splitting or destocking.

Valo's young wife, Seju, also does not wish to travel in the bush with the camp. They have abandoned the camel as a draught animal and begun migrating with a van to make the journey more comfortable and improve their social status, affecting the socio-temporal organization of their mobility as well as the group structure. In the longer term, Seju wishes to remain in the village to educate her children so that they may access alternative livelihood opportunities. Given these circumstances and uncertainties for the future, Nathubhai remains dismayed at the failure of generational succession for his livelihood and flock.

Therefore, while the Rabari display immense flexibility and adaptability in the face of uncertainties, we must ask: Why must pastoralists always adapt, and for how long? Why must they bear the burden of unfavourable policies rather than demand safeguards and enabling environments? When options for flexible adaptation are increasingly constrained and difficult to realize, pastoralists' agency is reduced in the face of political-economic forces outside their control. Hence, there is a thin line between highlighting adaptability even under challenging circumstances and becoming insensitive to the slow violence of accelerated transformations under capitalism (Fitz-Henry, 2017).

Therefore, not only is it crucial to recognize uncertain circumstances and the inherently ambiguous, flexible, and uncertain strategies and structures of the pastoralists, but it is also important to draw from pastoralism the more fundamental challenges to the singular views of progress, modernity, and development that underpin restrictive policies. There are always multiple development trajectories, and avoiding the invisible foreclosing of possible, alternative futures must be prioritized.

Instead, embracing uncertainties as a framework offers a more plural vision of progress with multiple versions of modernity that incorporate different viewpoints and pathways (cf. Scoones and Stirling, 2020). It disrupts the binaries that pitch pastoralists in opposition to progress and development. It recognizes the contemporaneity of pastoral strategies and capacities and so allows for the reclaiming of the pastoral present.

References

Adams, V., Murphy, M. and Clarke, A.E. (2009) 'Anticipation: technoscience, life, affect, temporality', *Subjectivity* 28: 246–65.

Bauman, Z. (2000) *Liquid Modernity*, Polity Press, Cambridge.

Bharwada, C. and Mahajan, V. (2006) 'Quiet transfer of commons', *Economic and Political Weekly* 41: 313–15.

Burges, J. and Elias, A. (eds) (2016) *Time: A Vocabulary of the Present*, New York University Press, New York NY.

Butt, B. (2016) 'Ecology, mobility and labour: dynamic pastoral herd management in an uncertain world', *Revue Scientifique et Technique de l'OIE* 35: 461–72.

FAO (Food and Agriculture Organization) (2021) *Pastoralism: Making Variability Work*, FAO Animal Production and Health Paper 185, FAO, Rome.

FAO (2022) *Making Way: Developing National Legal and Policy Frameworks for Pastoral Mobility*, FAO Animal Production and Health Guidelines 28, FAO, Rome.

Fitz-Henry, E. (2017) 'Multiple temporalities and the nonhuman other', *Environmental Humanities* 9: 1–17 <https://doi.org/10.1215/22011919-3829109>.

Guyer, J.I. (2007) 'Prophecy and the near future: thoughts on macroeconomic, evangelical, and punctuated time', *American Ethnologist* 34: 409–21 <https://doi.org/10.1525/ae.2007.34.3.409>.

Hazan, H. and Hertzog, E. (eds) (2011) *Serendipity in Anthropological Research: The Nomadic Turn*, Ashgate, Farnham.

Ingold, T. (2000) *The Perception of the Environment: Essays on Livelihood, Dwelling and Skill*, Routledge, London.

Jacques, E. (1982/1990) 'The enigma of time', in J. Hassard (ed.), *The Sociology of Time*, pp. 21–34, Palgrave Macmillan, New York.

Kaufmann, J.C. (2009) 'The sediment of nomadism', *History in Africa* 36: 235–64.

Kolinjivadi, V., Vela Almeida, D. and Martineau, J. (2020) 'Can the planet be saved in *Time*? On the temporalities of socionature, the clock and the limits debate', *Environment and Planning E: Nature and Space* 3: 904–26 <https://doi.org/10.1177/2514848619891874>.

Krätli, S. and Schareika, N. (2010) 'Living *off* uncertainty: the intelligent animal production of dryland pastoralists', *European Journal of Development Research* 22: 605–22 <https://doi.org/10.1057/ejdr.2010.41>.

Maffesoli, M. (1998) 'Presentism – or the value of the cycle', in S. Lash, A. Quick and R. Roberts (eds), *Time and Value*, pp. 103–12, Blackwell, Oxford.

Mehta, L. (2005) *The Politics and Poetics of Water: The Naturalisation of Scarcity in Western India*, Orient Blackswan, Delhi.

Robbins, P. (2004) 'Pastoralists inside-out: the contradictory conceptual geography of Rajasthan's Raika', *Nomadic Peoples* 8: 136–49 <https://doi.org/10.3167/082279404780446032>.

Roe, E. (2020) 'A new policy narrative for pastoralism? Pastoralists as reliability professionals and pastoralist systems as infrastructure', STEPS Working Paper 113, STEPS Centre, Brighton.

Rosa, H. (2013). *Social Acceleration: A New Theory of Modernity* (J. Trejo-Mathys, Trans.). Columbia University Press, New York.

Scoones, I. and Stirling, A. (eds) (2020) *The Politics of Uncertainty: Challenges of Transformation*, Routledge, London <https://doi.org/10.4324/9781003023845>.

TNN (2020) 'Livestock export from Tuna port gets govt nod', *Times of India*, 30 April. Available from: <https://timesofindia.indiatimes.com/city/rajkot/livestock-export-from-tuna-port-gets-govt-nod/articleshow/75458580.cms>.

Westphal-Hellbusch, S. and Westphal, H. (1974) *Hinduistische Viehzüchter im Nord-westlichen Indien: Die Rabari*, vol. 1, Duncker and Humblot, Berlin.

CHAPTER 4
Hybrid rangeland governance: ways of living with and from uncertainty in pastoral Amdo Tibet, China

Palden Tsering

Introduction

Ecological, social, political, and economic variabilities combine in the pastoral areas of Amdo Tibet in China, bringing with them multiple uncertainties. Finding ways to respond to these uncertainties is central to pastoralists' successful use of rangelands. During my fieldwork, I would ask people, 'How do you understand uncertainty?' I can still recall the first time I posed this question when visiting the winter pastures in the southern part of Kokonor in November 2018. In a small winter house, while sipping milk tea, the host, Suby, a male pastoralist with a white baseball cap, sunglasses, and ruddy cheeks, said calmly, 'Why worry about uncertainty? We cannot foretell anything, so why bother to worry about tomorrow? Yesterday is already the past, and we do everything we can to address the issues of today.'

Like Suby, pastoralists across Amdo Tibet have long faced highly variable conditions such as tribal disputes, natural disasters, climate change, and policy shifts. In other words, pastoralists have always lived with uncertainties; to them, uncertainty is the only certainty. Such collective world-views and beliefs about uncertainty are in turn reflected in livelihood and management strategies. This is why it was important to understand uncertainty through the eyes of pastoralists. From 2019 to 2022, I conducted in-depth interviews, focus group discussions, participant observation, and photovoice exercises in Saga and Lumu villages, two pastoral settings in Amdo Tibet, China (Figure 4.1).

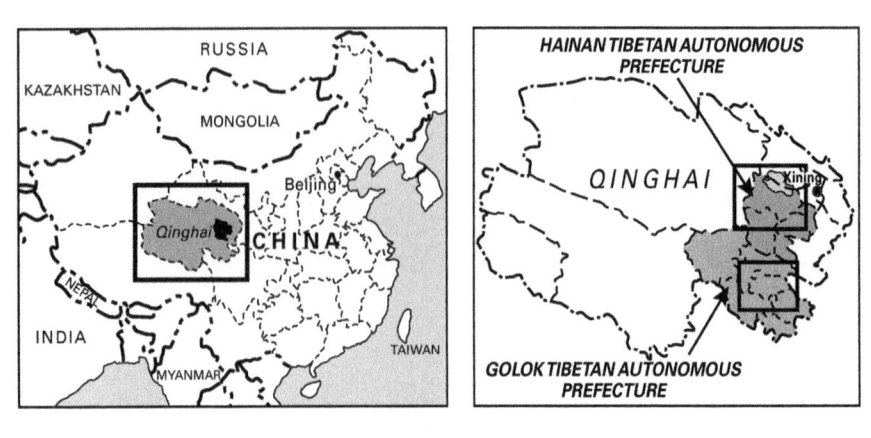

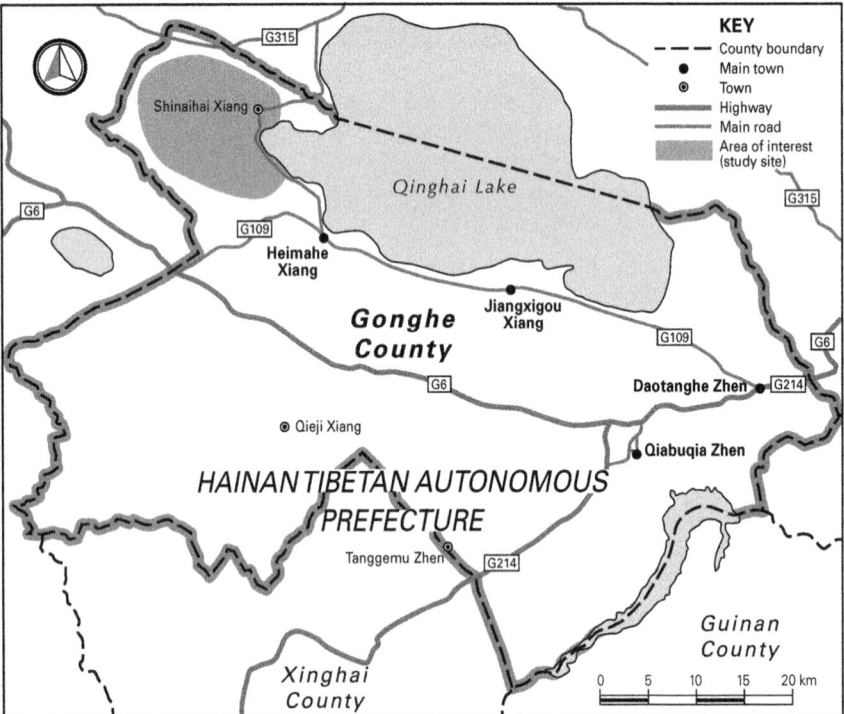

Figure 4.1 Saga and Lumu in Amdo Tibet, China

Seeing uncertainty in pastoral Amdo Tibet

How different groups of people construct knowledge through constant engagement with the dynamic world is the key to exploring the understandings and implications of uncertainty in the Tibetan context. In other words, pastoralists' knowledge not only reflects their beliefs and world-views but also actively shapes the changing world.

The uncertainties that pastoralists are most focused on are visible, traceable, everyday uncertainties, where people have the capacities to make a difference. During the photovoice sessions, pastoralists rarely identified political and policy shifts as sources of uncertainty.[1] This does not mean that pastoralists are unaware of such uncertainties: they all experience the consequences of major policy changes, resource grabbing, and ecological resettlement. They opt instead to disregard these uncertainties because they are outside their control; their voices are not heard and they have limited agency in relation to externally imposed policies.

In Tibetan Buddhism studies, uncertainty and the closely related terms 'impermanence', 'emptiness', and 'mid-way' are widely explored (Thanissaro, 2012; Todd, 2015). I got the opportunity to visit a *Geshe* (དགེ་བཤེས། 格西, the title given to a Tibetan Bhuddist scholar) from Saga in his cave-like quarters on a lovely summer afternoon in August 2019. The *Geshe* fully comprehended my explanation of uncertainty – focused on the condition where we don't know the likelihood of certain outcomes (see Chapter 1) – but he found it difficult to identify an equivalent Tibetan term. After finishing his butter tea from a brown wooden bowl, the *Geshe* began to explain his interpretations: 'In Buddhism, we use *mi rtag pa*:[2] what happened is already in the past, and what is going to happen is unpredictable; all we can depend on is the present, we deal with what is happening now.'

The Buddhist perspective on uncertainty centres on the nature of the consequences. As indicated by the *Geshe*'s quote, things change due to their impermanence. Thus, it is vital to recognize that the ultimate consequence of everything is destructive in order to embrace the ongoing, perpetual, and contingent flow of processes and relations. In this perspective, uncertainty is the impermanent consequence of constant changes, and thus the interpretation stresses the realization, acceptance, and embracing of changing realities.

The expression *mi rtag pa* is widely used to denote that the condition of things is always changing and disappearing in each instant. The term emphasizes the impermanent and destructive aspect of existence, since nothing is permanent but only continuously changes. Nevertheless, as the *Geshe* explained, the understanding of uncertainty has other connotations: 'Uncertainty is everywhere, it is unpredictable, unimaginable, but unlike the concept of *mi rtag pa*, it has consequences, results, and particular impacts.'

After this discussion, the *Geshe* offered the term *nges med* (ངེས་མེད་འགྱུར་ལྡོག), which refers to the Buddhist concept of change in motion, the state of continuous change. *nges med*, unlike *mi rtag pa*, focuses on the continuity of change, its processes and relations. The state of society or the local ecosystem therefore has to be understood in relation to constantly changing processes – both past and present – as verbs, not nouns (Hertz et al., 2020). For instance, in the case of the expansion of the Kokonor lake, which is taking up important pasture and displacing pastoralists, the cause – whether climate change or expanding conservation efforts – is less of a concern compared to the consequences. As the *Geshe* stressed, *nges Med* does not provide a comprehensive understanding and so additional phrases are required to illustrate uncertainty.

Bsam yul las das pa is commonly used by both secular and non-secular Tibetans. It refers to things, events, and states that are beyond one's imagination, thought, and experience. If something is *bsam yul las das pa*, it pertains to things that are unpredictable and unmeasurable, a state of simply not knowing that cannot be prepared for.

Bsam, the Tibetan word for thought, refers to the capacity for thought or thinking ability. *Yul* signifies the location or object, which here refers to the object of the thought, whereas *das pa* refers to things beyond thought and capacity for thought. This phrase therefore implies the consequences of the ongoing and changing processes, with the emphasis on responses to unexpected outcomes. However, in this case, experiences are key and conditions of those experiences matter for possibilities of adaptation and transformation.

As West et al. (2020: 311) argue, our engagement with 'the social, material and technological aspects of holistic, unfolding situations produces experience'. As the *Geshe* concluded, *bsam yul las das pa* refers to the things and events beyond one's experience; thus, uncertainties such as snowstorms and wildlife attacks are not included, as they are more or less resolvable based on existing and accumulated experiences and practices, despite the fact that the forms and scale of these uncertainties vary over time and space. Pastoralists are able to respond to such uncertainties due to their accumulated individual and collective experiences, inherited over time across generations. Thus, *bsam yul las das pa* focuses on the realm of novelty, particularly things and events that are beyond one's experience, where no responses have been worked out before.

Through my discussions with the *Geshe* and others, I learned that there is no single, definitive, equivalent Tibetan term for 'uncertainty'; rather, there are multiple phrases and expressions that convey the term in different contexts. *Mi rtag pa* focuses on the impermanent consequences of things; *nges med* emphasizes the ongoing process, while *bsam yul las das pa* relates to the conditions of experience (Table 4.1).

Table 4.1 Different responses to uncertainty in the Tibetan context

Different context	Different foci	Different responses
Mi rtag pa	Impermanence	Staying with the changes/ vulnerability unmeasurable
Nges med	Ongoing process of change	Customizing rules and relations/ adaptabilities and transformability/ vulnerability depends on variables
Bsam yul las das pa	Conditions of experience	Experiences from the past matter

Uncertainty in the Tibetan context

However, in their day-to-day practice, pastoralists must continuously adapt to uncertainties, making use of a well-developed, practical repertoire of responses, even if the terms used are varied. In order to go beyond the philosophical debates around terms and phrases, I explored ideas of uncertainty in grounded contexts through two rounds of photovoice discussion (see Chapter 2).[3] In these exercises, pastoralists – both men and women in the two sites in Amdo – expressed their understandings of uncertainty through photographs and narrative interpretations. During a photovoice discussion in November 2019, TJ, a father of three children from Golok, shared his perspective, reflecting on a photograph of a senior couple who chose to stay with their livestock on the pastures.

> Pastoralism is the only thing we are good at. Most of the elders already moved to the town and close to the monasteries. And I think the further you get away from your grassland, the more uncertain your life gets. For example, the quality of the food you take is *mi rtag pa* (uncertainty); the milk powder you buy for your grandchildren is *mi rtag pa*; the fake butter and raw meat you purchase from the market is *mi rtag pa*. You are not sure (*mi rtag pa*) about the prices of the electricity and gas that you pay monthly for the heat because you don't collect yak dung anymore. You are not sure (*mi rtag pa*) about your wellbeing because you don't share stories and you don't have temples and stupas to circulate. Therefore, the only certain thing for us is pastoralism, even though there are uncertain occasions like heavy snowfall and wildlife attacks. But compared to the urban and non-pastoral way of life, our life is more certain in all different ways because we are familiar with and we know how to handle things.

Uncertainty, according to TJ, consists of unexpected relationships and interactions with the environment, which is why his familiar lifestyle inspires confidence. In the same discussion, TJ emphasized that, 'According to Tibetan Buddhism, all existence is impermanent, and uncertainty is one of its aspects. It does not imply that we should give up because everything eventually comes to an end. Realizing the fundamental nature of existence is the essence of impermanence and uncertainty, and this understanding enables us to make decisions throughout the various periods of life when confronted with uncertainties.'

Another reflection on uncertainty comes from Lhamo, a mother of two sons. During a group discussion in Saga, Kokonor, in January 2020, she noted:

> One of the many *mi rtag pa* (uncertainties), and the one most pertinent to pastoralism, is the loss of pasture due to lake expansion. The loss made life difficult for my family and me. I was born on this property as a pastoralist and grew up herding the flock and consuming the milk and meat of the animals. My primary source of family income is the sale of animals and dairy products, such as butter and cheese. Due to the loss

of land, it is now extremely difficult to earn a living. We must now rent pasture from relatives and friends, and these pastures are frequently of lower quality and further away. Therefore, renting pasture and herding livestock now necessitates greater investments in land, labour, and transportation.

However, we all know that this [lake expansion and land loss] is not going to be permanent; the lake expands and also shrinks. Moreover, we started to do eco-tourism here, my sons have started a home-stay centre here for the tourists, and it is good money, especially during summer time. So, this *Mi Rtag Pa* will also change, and we will have our pasture back.

Uncertainty, according to Lhamo, is not something permanent; rather, it is central to the fluidity of things. Uncertainty brings challenges but also opportunities for transformations; in her case, gaining profit from engaging in eco-tourism on the winter pasture. Thus, in the Tibetan context, the essence of uncertainty is centred on fluidity, something that is always in motion and is focused on the immediate challenges of things that are 'taking place'. Uncertainty is therefore the endless combination of dynamic processes and relations from various actors and realms, whether material or immaterial, and so central to pastoral livelihoods.

The fluid processes and connections at the centre of interactions between nature and humans enable pastoralists both to live with and from uncertainty, making use of uncertainties as possibilities and opportunities for adaptation and transformation (Mancilla García et al., 2020). Thus, embracing uncertainty means focusing on interactions within and between ongoing complex adaptive systems (Preiser et al., 2018). The question then becomes how Amdo pastoralists transform these perceptions into actions on the ground.

Living with and from uncertainty

Increasing uncertainties pose considerable challenges for governing rangelands. Uncertainties include climate change, new large-scale infrastructural investments, the establishment of new settlements, resource grabbing, and land fragmentation. All these combine to increase the challenges for pastoral livelihoods. Navigating uncertainties, responding to the negative effects, and making use of opportunities requires considerable skill and has resulted in a series of practical innovations around rangeland governance: the management of land used by pastoral livestock.

Pastoralists' strategies must also confront state policies on rangeland governance, which add another layer of uncertainty. In recent decades, these have taken a market-oriented stance linked to policies for 'rural vitalization', 'ecological civilization', and the reinforcing of private property rights. Through these policies, the state has provided incentives for changing land use on the rangelands. However, these changes may not help with navigating diverse uncertainties and sustaining pastoral livelihoods, so pastoralists always have to seek compromises between local practices and state impositions.

State policies invariably assume a stable, regulated form of property and land use, with clearly defined patterns of tenure. On the ground, however, pastoralists must maintain more flexible, negotiated, hybrid arrangements to live with and from diverse uncertainties. The disconnection between what state policies and what exists on the ground is sometimes the basis for contention and conflict. However, the state's ability to impose and enforce is always limited, and compromises emerge. This results in a hybrid constellation of roles, rules, and relationships, which allow pastoralists to navigate uncertainties effectively (Scott, 1990; Simula et al., 2020; Tsering, 2022a).

As a result, various arrangements around land use are improvised, reformulated, and remade from below. These do not necessarily directly challenge state power (e.g., Hall et al., 2015; Ptáčková, 2019); rather, pastoralists try to find ways to negotiate, seeking practical solutions that work in local contexts. These emerge through what I call 'practices of assemblage' (Tsering, 2022a, b; Li, 2014), where local practices emerge in response to uncertain contexts (Ho, 2005, 2017; Heilmann and Perry, 2011; Yeh et al., 2013; Scoones et al., 2018). As the cases below show, assemblage is the technique of drawing heterogeneous elements together, forging connections between them, and sustaining these relationships in the face of uncertainty (Li 2014, 2021). Rangeland governance is therefore not restricted to formal edicts from the state but is centred on the fluid practices, negotiations, and contestations involving multiple resource user groups on the ground.

The following two sections explore such hybrid rangeland governance in Amdo Tibet, highlighting how uncertainties are embraced through performative practices situated in context, and how responses are assembled through a variety of means.

The role of the monasteries in resource governance

In pastoral Lumu in Golok, the role of the monastery is pervasive in everyday social, cultural, and religious life. Despite the central state's power, there is a need to combine diverse forms of authority, and this results in the emergence of negotiations around highly contentious state policies. The case of the construction of a mineral water factory on a private winter pasture in Lumu showed how pastoralists, or 'rangeland contractors'[4] often approached the monastery for advice.

As Uncle Bam a contractor on the land, reflected when interviewed in February 2021:

> I went to the monastery and consulted the incarnate, the incarnate recommended not to give the permit [to the water factory company] because a mineral water factory will bring damage to the water, not only the water we drink, but also the water that our livestock, the wildlife, and the downstream villagers use.

Figure 4.2 Members of the local monastery carrying hay for the blue sheep in Golok.
Credit: Palden Tsering

In Lumu, the monastery serves as both the local authority and intermediator between the villages and local government. It takes these roles seriously and, consequently, the de facto use and access to resources is continuously negotiated and contested with the participation of various resource users (Tsering, 2019; Simula et al., 2020). In December 2019, Jab, the monastery secretary, concluded:

> The monastery is determined to serve all sentient beings. We are never opposed to development, we just worry about how to develop. This land belongs to the pastoralists, the monastery, livestock, and wildlife. Therefore, we cannot just give the land for a mineral water factory; this is selfish. We must think about others because all things are interconnected, and we must remember this.

This use of rangeland in Lumu is governed not only by land contractors but also their relationships with the bigger world, including the monastery. Land is viewed as pasture for livestock, as a habitat for wildlife, and as a sacred site for the monastery. The inclusion of the monastery is therefore crucial in any decision linked to rangelands and land use in the area.

The existence of the monastic network and the mediation role of religious leaders are essential for negotiation-based, hybrid rangeland management. The presence of the monastery also creates 'possibility spaces' (directly or indirectly) for participation and engagement of local pastoralists in policy processes, allowing them to claim their rights to land, sometimes confronting powerful external investors.

Pluralist resource governance

In Saga village, near the Kokonor lake area, pastoralists must work with many actors to strengthen their bargaining power and advance their own interests in the face of strong efforts to change land use by external actors, both state conservation agencies and private investors interested in the tourism potential of the area. Rather than direct confrontation, pastoralists challenge government policies through informal networks and more hidden forms of passive resistance (Scott, 1985, 1990). These forms of resistance can be seen from pastoralists' flexible ways of responding to different state projects. Interviewed in February 2021, Uncle Bam, the former village party secretary, explained:

> The government claimed that the village needed a grazing ban zone in order to receive the subsidy. Pastoralists wanted the subsidy and also worried that they would no longer be able to herd livestock. Therefore, the village decided to design the collectively used summer pasture as the grazing ban zone because it is far away from the government; therefore, it will be hard for inspections.

Subsidies are allocated to pastoralists if they agree to prioritize rangeland conservation over animal grazing, through designating their pasture as part of the grazing ban zone and decreasing animal numbers. This is in line with state policies for conserving rangeland due to ecological concerns about rangeland degradation and desertification. But, according to Uncle Bam, there were many local, collective concerns about rangeland loss due to government projects, such as the grazing ban zone policy. Instead of complying completely, pastoralists opted to maximize rangeland use by designating an area for the 'ban' that was used for only a few weeks, was far from the township centre, and was not subject to regular inspections, so could still be used surreptitiously. Meanwhile, pastoralists shifted their grazing to the winter pasture areas that were not being targeted by the state. This meant that pastoralists moderated the impacts of the state project by adjusting their use of rangeland areas, making use of their livestock management abilities to adapt.

Apa Tashi, a 58-year-old herder who lost half of his winter pasture to the lake, complained:

> They [the government] always choose quick-fix solutions. They relocated my family to the township centre right after the lake expansion. It is acceptable for me to move to the township, but as a pastoralist, my livelihood depends on livestock and rangeland. What is the good if I cannot make my belly full [living in the township without livestock and land]? It will only create issues for the government.

As Apa Tashi argued in this July 2020 interview, quick-fix solutions may resolve temporary issues (in his case, the loss of a winter house to the lake expansion), but further livelihood policies for the landless pastoralists need to be negotiated. Apa Tashi's access to the rangeland is critical to his capacity to continue his way of life. Numerous state-designed strategies for rangeland conservation and mitigation of lake expansion have consistently targeted the removal of pastoralists from their rangelands as a solution.

However, some pastoralists view policies such as the grazing ban and resettlement as advantageous. For some, the creation of the grazing ban zone on the summer pasture worked out well. In another interview in July 2020, Apa Libo, who sold all his livestock and moved to the township centre in 2015, told me:

> I am happy the village and the government [township level] designed the summer pasture as the grazing ban zone because I don't have any livestock; thus, summer pasture is useless to me, and now there is money [the grazing ban subsidy], which is great for those who don't graze on the summer pasture.

Others with few animals move their stock to the winter pasture and so also do not worry about the ban.

In this case, for those other than some larger livestock owners, the negotiated solution worked out well. The hybrid arrangement, a result of informal negotiation and centred on hidden forms of compromise agreement, provided a solution whereby pastoralists both received a subsidy and could continue to graze their animals. The intermediation of local government officials at village and township level helped with the negotiation, and for now at least, a satisfactory outcome has been brokered.

Negotiating solutions for adaptive responses

Both cases show how uncertainties emerging from state policies and competition over resources from multiple actors (exacerbated by climate change in the case of the lake expansion) were addressed by creating new, brokered solutions. The process of assemblage, responding to the unfolding and fluid nature of each context, resulted in new elements being composed, with new roles, rules, and relationships constantly emerging.

As state policies shift, so must the local practices of assemblage and the form of hybridity in land governance that emerges at a local level in response to uncertainty. Collaborative approaches to assemblage involving multiple actors result in higher participation from resource users, greater investment in appraisal before the design and implementation of interventions, and improved conflict negotiation mechanisms. All these elements are important for the adaptive management of resources and the building of resilience under uncertain conditions (Berkes et al., 2000; Folke, 2006, 2016).

This has practical implications for how resource management is designed and practised in rangeland areas such as Amdo Tibet. If uncertainties are to be embraced, and lived with and from, authorities must devote additional time and effort to becoming informed about local contexts during policy design, formulation, and implementation. Informed policymaking stresses a thorough understanding of local politics, power dynamics, power relations, and grounded practices. If processes of assemblage are to be facilitated, multilateral and multilayered processes involving multiple actors always need to be encouraged.

In the Amdo Tibet case, this means that local government officials (at the prefectural, county, and township level), local institutions (the monastery, nunnery, monastic associations, grass-roots civic organizations), and pastoralists themselves (as landholders and users) all need to engage in the formulation, reformulation, and implementation of policies. This means a shift away from a centralized, fixed, and dominant style of policymaking towards an inclusive, co-produced policy process that stimulates inclusion and innovation.

An informed approach to policymaking thus means not just seeing through the eyes of pastoralists but also thinking as they do, grasping how 'uncertainty' is understood through local, culturally embedded perceptions, as discussed earlier. In this way, with common understandings, a hybrid approach may facilitate interactions, negotiations, and bargaining between informal and formal authorities over policy outcomes.

Pastoralists are not necessarily opposing a particular policy or intervention; rather, they challenge top-down designs and try to incorporate new initiatives from the government into their local repertoires. They must adapt continuously, tailoring materials at hand through customizing roles, enlisting different players into their networks – such as the monastery or local government officials, for example – through various practices of assemblage, and so constructing new rules for resource management in a plural, highly fluid context.

In this way, what I call hybrid governance of rangelands emerges. As a route to more effective cooperation, mutual understanding, and co-production amongst diverse actors, the process offers opportunities for the state to see plans emerge more effectively, while for pastoralists, options are more likely to be embedded in existing practices, allowing them to continue their production and respond flexibly to uncertainty.

Conclusion

There is no single, definitive Tibetan term for uncertainty; rather, different phrases convey the concept in various circumstances. The core of uncertainty in the Tibetan context is fluidity, something that is constantly in motion and constituted through things that are 'happening'. Uncertainty is the unending mixing of dynamic processes and relationships between diverse material and immaterial actors and domains. The interdependent relationship between humans and nature enables pastoralists to live with and from uncertainty, utilizing uncertainties as possibility spaces for adaptation and transformation.

Through multi-case ethnographic research and a mixed method approach in two pastoral villages in Amdo Tibet, China, findings reveal that ongoing rangeland governance practices are constructed through different practices of assemblage. The result is a hybrid regime that goes beyond the classic description of private, common, or state-led forms of tenure. A hybrid approach – and especially the process of building assemblages of actors, practices, technologies, and forms of knowledge – in turn allows herders both to respond to uncertainties as they arise, as well as make the most of opportunities that emerge from uncertain settings.

Notes

1. See seeingpastoralism.org.
2. མི་རྟག་པ, impermanence, refers to the state of things that are changing and disappearing each instant.
3. Caroline Wang and Mary Ann Burris defined photovoice in the early 1990s as 'a participatory action research method, from the photos that the participants took, people can identify, represent, and enhance their community through a specific photographic technique' (Wang and Burris, 1997; Bennett and Dearden, 2013, quoted in Apaza and DeSantis (2011: 369)). See more on the PASTRES Photovoice website, seeingpastoralism.org.
4. This specifically refers to the holders of a rangeland household contract.

References

Apaza, V. and DeSantis, P. (2011) *Facilitator's Toolkit for a Photovoice Project*, United for Prevention in Passaic County and the William Paterson University Department of Public Health, the William Paterson University Print Services, USA.

Bennett, N.J. and Dearden, P. (2013) 'A picture of change: using photovoice to explore social and environmental change in coastal communities on the Andaman coast of Thailand', *Local Environment* 18: 983–1001 <https://doi.org/10.1080/13549839.2012.748733>.

Berkes, F., Colding, J. and Folke, C. (2000) 'Rediscovery of traditional ecological knowledge as adaptive management', *Ecological Applications* 10: 1251–62 <https://doi.org/10.2307/2641280>.

Folke, C. (2006) 'Resilience: the emergence of a perspective for social–ecological systems analyses', *Global Environmental Change* 16: 253–67 <https://doi.org/10.1016/j.gloenvcha.2006.04.002>.

Folke, C. (2016) 'Resilience (Republished)', *Ecology and Society* 21: 44 <https://doi.org/10.5751/ES-09088-210444>.

Hall, R., Edelman, M., Borras, Jr, S.M., Scoones, I., White, B. and Wolford, W. (2015) 'Resistance, acquiescence or incorporation? An introduction to land grabbing and political reactions "from below" ', *Journal of Peasant Studies* 42: 467–88 <https://doi.org/10.1080/03066150.2015.1036746>.

Heilmann, S. and Perry, E.J. (eds) (2011) *Mao's Invisible Hand: The Political Foundations of Adaptive Governance in China*, Harvard University Asia Center, Cambridge MA.

Hertz, T., Mancilla García, M. and Schlüter, M. (2020) 'From nouns to verbs: how process ontologies enhance our understanding of social-ecological systems understood as complex adaptive systems', *People and Nature* 2: 328–38 <https://doi.org/10.1002/pan3.10079>.

Ho, P. (2005) *Institutions in Transition: Land Ownership, Property Rights, and Social Conflict in China*, Oxford University Press, Oxford <https://doi.org/10.1093/019928069X.001.0001>.

Ho, P. (2017) *Unmaking China's Development: The Function and Credibility of Institutions*, Cambridge University Press, Cambridge <https://doi.org/10.1017/9781316145616>.

Li, T.M. (2014) 'What is land? Assembling a resource for global investment', *Transactions of the Institute of British Geographers* 39: 589–602 <https://doi.org/10.1111/tran.12065>.

Li, T.M. (2021) 'Commons, Co-Ops, and Corporations: Assembling Indonesia's Twenty-First Century Land Reform', *Journal of Peasant Studies* 48: 613–39 <https://doi.org/10.1080/03066150.2021.1890718>.

Mancilla García, M., Hertz, T., Schlüter, M., Preiser, R. and Woermann, M. (2020) 'Adopting process-relational perspectives to tackle the challenges of social-ecological systems research', *Ecology and Society* 25: 29 <https://doi.org/10.5751/ES-11425-250129>.

Preiser, R., Biggs, R., De Vos, A. and Folke, C. (2018) 'Social-ecological systems as complex adaptive systems: organizing principles for advancing research methods and approaches', *Ecology and Society* 23: 46 <https://doi.org/10.5751/ES-10558-230446>.

Ptáčková, J. (2019) 'Traditionalization as a response to state-induced development in rural Tibetan areas of Qinghai, PRC', *Central Asian Survey* 38: 417–31 <https://doi.org/10.1080/02634937.2019.1635990>.

Scoones, I., Edelman, M., Borras, Jr., S.M., Hall, R., Wolford, W. and White, B. (2018) 'Emancipatory rural politics: confronting authoritarian populism', *Journal of Peasant Studies* 45: 1–20 <https://doi.org/10.1080/03066150.2017.1339693>.

Scott, J.C. (1985) *Weapons of the Weak: Everyday Forms of Peasant Resistance*, Yale University Press, New Haven CT.

Scott, J.C. (1990) *Domination and the Arts of Resistance: Hidden Transcripts*, Yale University Press, New Haven CT.

Simula, G. et al. (2020) 'COVID-19 and pastoralism: reflections from three continents', *Journal of Peasant Studies* 48: 48–72 <https://doi.org/10.1080/03066150.2020.1808969>.

Thanissaro, P.N. (2012) 'What makes you *not* a Buddhist?: A preliminary mapping of values', *Contemporary Buddhism* 13: 321–47 <https://doi.org/10.1080/14639947.2012.716709>.

Todd, S. (2015) 'Experiencing change, encountering the unknown: an education in "negative capability" in light of Buddhism and Levinas', *Journal of Philosophy of Education* 49: 240–54 <https://doi.org/10.1111/1467-9752.12139>.

Tsering, P. (2019) 'The role of Tibetan monastic organizations in conservation and development: the case study of Shar 'od monastery in Golok, China', *Journal of Tibetology* 24: 276–89.

Tsering, P. (2022a) 'Institutional Hybridity: Rangeland Governance in Amdo, Tibet', doctoral dissertation, University of Sussex. Available from: <https://core.ac.uk/download/pdf/519857982.pdf>.

Tsering, P. (2022b) 'Jarmila Ptáčková, exile from the grasslands: Tibetan herders and Chinese development projects', *Nomadic Peoples* 26: 282–86 <https://doi.org/10.3197/np.2022.260207>.

Wang, C. and Burris, M.A. (1997) 'Photovoice: concept, methodology, and use for participatory needs assessment', *Health Education and Behavior* 24: 369–87 <https://doi.org/10.1177/109019819702400309>.

West, S., Haider, L.J., Stålhammar, S. and Woroniecki, S. (2020) 'A relational turn for sustainability science? Relational thinking, leverage points and transformations', *Ecosystems and People* 16: 304–25 <https://doi.org/10.1080/26395916.2020.1814417>.

Yeh, E.T., O'Brien, K.J. and Ye, J. (2013) 'Rural politics in contemporary China', *Journal of Peasant Studies* 40: 915–28 <https://doi.org/10.1080/03066150.2013.866097>.

CHAPTER 5
Uncertainty, markets, and pastoralism in Sardinia, Italy

Giulia Simula

Introduction

In Sardinia in Italy, pastoralists constitute the great majority of those who work in agriculture. (Sardegna Agricoltura, 2013). The most important pastoral products are sheep milk and cheese, and the quantities produced stand out in both the Italian and European markets, with Sardinia as the Italian region where most sheep cheese is produced and a leading region in Europe (Ismea, 2019).

One can see that Sardinia is a land of pastoralists, not only by looking at the data on milk production but also by looking at the history, the landscape, and the culture. Sardinian songs often tell stories of transhumance, of migration, of struggles over land, and the tyranny of rich landowners. Pastoralism in Sardinia is a very important part of how people build their own livelihood; it is often mixed with other activities and, as shown in Table 5.1, there is a high percentage (almost 30 per cent) of very small farms. There are around 12,000 sheep farms in Sardinia and around 70 per cent of the farms have less than 300 sheep.

Across this range of farm and flock sizes, there are different types of production, ranging from highly extensive (including some who still practise transhumance) to those who have intensified in a fixed farm site, with high levels of inputs and mechanization, much of the feed supplied through forage crops grown on site, and the use of purchased supplements. Each pastoralist must navigate many uncertainties – from a changing climate to highly variable market conditions. Livelihoods and therefore production systems are constructed around such adaptive responses to uncertainty. In some cases, pastoralists must live with and from uncertainty, responding flexibly to changing conditions, including developing high-value niche products, such

Table 5.1 Distribution of sheep farms by flock size

Flock size	Number of sheep farms	% of total sheep farms	Number of sheep	% of total sheep	Average flock size
<100	3,455	28.65	148,119	4.88	43
100–300	4,982	41.32	976,736	32.16	196
300–500	2,245	18.62	860,538	28.34	383
500–700	778	6.45	455,775	15.01	586
>700	598	4.96	595,598	19.61	996
Total/Average	**12,058**		**3,036,766**		**252**

Source: Atzori et al. (n.d.: 8).

as artisanal cheeses, or diversifying to activities outside livestock production. Others attempt to tame uncertainty by investing in production systems that aim to control and manage risk and reduce variability. Depending on the asset base, networks and connections, and people's attitudes and outlooks, very different styles of pastoralism exist in Sardinia. There is no one single pastoralism, but many (Simula, 2022).

Where does the sheep milk go? The milk has different destinations. Some sell milk to cooperative dairies, managed and owned by pastoralists themselves. These cooperatives industrially process the milk into Pecorino Romano cheese. This cheese has the biggest market share and is mostly sold in the USA and elsewhere in Europe. While the cooperatives process and sell other types of cheese, the market is dominated by Pecorino Romano, with important consequences for pastoral livelihoods, as the chapter will show. Others are not part of a cooperative but rather sell the milk to private industries. There are also those who own mini-dairies to make artisanal cheese and sell directly to consumers in their shops. Finally, there are those who sell locally and informally to members of the community, focusing mostly on home consumption, and who complement this with other paid work.

While high-quality artisanal cheeses are produced for niche markets in Sardinia and abroad and sold to tourists who visit the island in huge numbers during the summer, the industrial cheese – Pecorino Romano – dominates the market. This results in a high dependence on the market conditions and prices of Pecorino Romano, which in turn has a great influence on the price of milk (Farinella and Simula, 2021). Given the variabilities of these markets, which are outside the control of pastoralists, this has created great uncertainties for them in the last decades.

Common uncertainties, different experiences, contrasting reactions

Sardinia has a typical Mediterranean climate: temperate with dry summers. The last decade has been characterized by more extreme weather events, such as droughts and floods, and added to this is the problem of recurrent fires during the summer.

Access to land and natural resources is key in the life of any rural producer. Privatization of land has drastically changed the pastoral landscape in Sardinia: land reform in the 1970s redistributed land and resulted in the semi-sedentarization of many pastoralists. In the mountainous areas of Sardinia, some common land areas remained but these, together with pastoralists' livelihoods, have been threatened by environmental conservation projects. Nowadays, land is also a great source of passive income thanks to European Union farm subsidies, and this leads to high levels of land speculation (Mencini, 2021).

The European Common Agricultural Policy (CAP) is one of the most important frameworks that influence the direction of economic incentives and policies in agriculture, livestock production, and pastoralism across the European Union. Uncertainty related to policies and institutions has several aspects. Policy frameworks change at every mandate with no long-term consistency. In some years, the production of wheat and cereals is incentivized, then it is disincentivized. The same is true for wine growing – first incentivized as a source of diversification and then disincentivized. And then it was the turn of sunflowers and cannabis. The CAP has created a lot of unexpected consequences, distortions, and reasons for speculation, and has thereby increased uncertainty in the life of pastoralists who, once adversely integrated into the formal sector and into international commodity chains, have become partly dependent on public incentives.

As already mentioned, markets are certainly among the biggest source of uncertainty for Sardinian pastoralists. Artisanal produce is often sold directly to consumers (with a maximum of two intermediaries) in local, national, and regional markets. What I call 'industrial cheese' is sold in international, European, and Italian markets. Industrial cheeses are linked to industrial processing and a standardized type of production. Throughout the last decades, more and more pastoralists have been incorporated into international markets linked to the booming production of Pecorino Romano cheese. While some increased their milk production and specialized in milk sales to the industries, others partially delinked from this commodity chain and produced cheese to sell locally. Market uncertainty in Sardinia is characterized by fluctuating and overall falling prices. This has resulted in the accumulation of profits and power in the input and distribution nodes of the milk commodity chain (Farinella, 2018).

All these uncertainties are widely recognized. But the solution offered by state officials and many experts is to address uncertainty as risk and attempt to predict and control it. 'Traditional' pastoralists are seen as 'backward' and in need of modernization. Only with developments such as consolidating farms, mechanizing production, and industrializing will pastoralists be able to benefit, it is argued. The narrative is that they need to be incorporated into markets and become more competitive and efficient.

The developmental narrative: trying to control uncertainty with efficiency and productivity

The state apparatus, mainstream research, and development agencies try to eliminate uncertainty based on the assumption that, with the right planning, pastoralists can produce more, more efficiently, and so increase their income. In this highly variable and uncertain context, research institutions, politicians, and experts who implement state policies try to manage and respond to the crises affecting the pastoral sector as if every condition were stable and predictable. This assumption is based on a vision of development and modernization that sees pastoralists incorporated into global commodity markets, one that responds to market crises with increasing efficiency, productivism, and profit.

For example, as a response to the milk price crisis – when milk prices crashed due to overproduction and lack of sales, and pastoralists came onto the streets to protest (Simula, 2019) – experts and politicians called for pastoralists to expand investments in their farms and to increase efficiency by reducing costs of production for each litre of milk. Agronomists and experts in animal nutrition, often working for multinational feed companies, suggested keeping track of the amount of milk produced by each animal and eliminating all the animals that were less productive to avoid wasting resources such as feed and fodder. Agricultural technology is also developing in this direction and the latest models of milking machines can track exactly the amount of food given to each animal and the amount of milk produced by each sheep.

This view emerges from several assumptions. The first is that the current crisis results from a lack of efficiency rather than from a problem of distribution of resources and profits in the commodity chain. The second is that pastoral farms function with an imperative of milk productivity and economic efficiency. The third is that pastoralists live in stable conditions and can therefore plan for high capital investments with predictable returns.

However, believing that such technical, productivist solutions can address the intersecting uncertainties created by markets, climate, and agricultural policy in the context of Sardinia is a sign that the state/expert perspective is far from the realities of pastoralists.

Pastoralists living with uncertainty

In contrast to the assumptions of many policymakers and experts, pastoralists do not live in stable conditions but in highly variable, uncertain, and often harsh and precarious situations. In this context, less productive animals might be more resistant to hostile territories and to diseases; they are important to maintain the flock biodiversity and to increase the flock's resistance to shocks and stresses. The ability to thrive in harsh environments with low pasture availability is a very important characteristic, according to pastoralists, who have a focus beyond an animal's efficiency in strictly productive terms.

Pastoral farms function within a complex system that is influenced by many elements, so assuming that economic efficiency – and an economic rationality based on a linear understanding of demand and supply and cost and benefit – is the guiding principle is a deep misunderstanding.

Rather than living in stable, controlled conditions with abundant availability of feed and fodder, most pastoralists work and live in variable and often precarious situations, with limited availability of natural and financial resources. They build their livelihoods taking advantage of the limited resources they have. As a result, pastoralists are very sceptical about top-down programmes and incentives. This is not because they are ignorant or 'backward', as they are very often portrayed, but because they know very well that they live and survive in uncertain circumstances. They necessarily work with contingency, always leaving several doors open as there are always multiple futures possible depending on what uncertainties impinge on them. They know that investing €200,000 (US$ 216,600) in an electric milking machine would mean entering a technological/production trap, unless they are extremely rich and have other income-earning options. To repay such a large debt and capital investment, they would need to make very high and consistent returns – an unlikely prospect for the vast majority of pastoralists in Sardinia living with predominantly low and very uncertain milk prices.

Although there are common uncertainties – from the market, climate, and policy environment, for example – these are lived, perceived, and experienced differently by pastoralists who inhabit different class positions, are of different genders, have different access to natural resources, and are different in respect of age and identity.

In this chapter, I will contrast two opposing realities of pastoralism in Sardinia, located in two different settings (Figure 5.1). On one side, I will look at the story of a livestock producer who engages in semi-intensive production in the plains area in the south of the island and sells milk to a private industry operator. In this case, the pastoralist follows the adage 'If you plan ahead, there is no uncertainty'; by investing and intensifying, uncertainties can be controlled. On the other side, I tell the story of a small-scale pastoralist living in the north of the island who can flexibly respond to uncertainty through a range of adaptive practices that generate reliability in the face of high levels of variability and therefore uncertainty.

Of course, there are many other types of pastoralism across a wide spectrum (see Simula, 2022), but these two examples highlight the contrast between developmental narratives of techno-managerial control and pastoralists' flexible practices of navigating uncertainty.

Managing uncertainty by trying to control it: the transformation of pastoralism into semi-intensive livestock production

When I met Tonino, one of my first questions was how long he had been a pastoralist. The fact that I called him a pastoralist made him laugh because

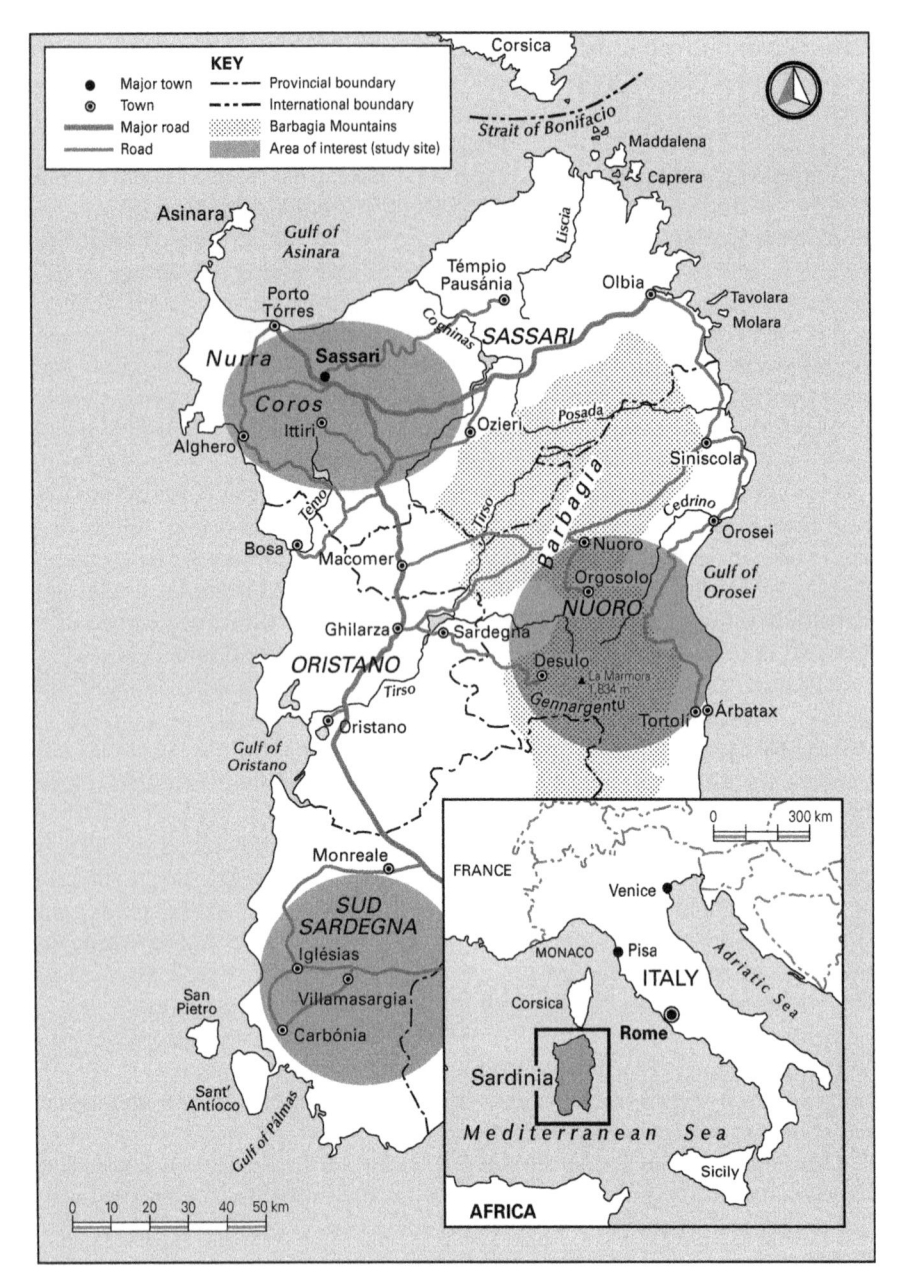

Figure 5.1 Map of Sardinia

according to him, that word is used for another category of producers, not him. He self-identifies as a livestock producer, not a pastoralist. Pastoralists, in his understanding, are all those producers that do not take this job seriously.

His origins are in Desulo, a town in the mountainous area of Sardinia. Both of his parents were from Desulo and his father was a transhumant pastoralist who moved to the plains of Villamassargia in the winter and then back to Desulo around May and until September. His father started buying land when he was still transhumant. At that time, the dairy industry in Sardinia was growing, and pastoralists were able to make a good income and started investing. Also, starting from the end of the 1960s, a process of state-led sedentarization and regional land reform encouraged his father to settle permanently in Villamassargia and to stop using transhumance.

The transformation of pastoralism into semi-intensive and sedentary livestock production is very much linked to state-led incentives and the process of industrialization. Tonino feels part of another generation; he supports the linear, evolutionary narrative according to which pastoralists have 'developed' into entrepreneurs and this is how he identifies. Nonetheless, he also has a strong sense of belonging to Desulo and he is proud of the pastoral culture and tradition where his roots are.

Tonino lives with his wife Marianna, who is also from Desulo. The family has its own vegetable garden and diversifies by producing and selling olive oil. Marianna does all the work in the house (caring for their son, cooking, and cleaning) and she also works in the olive grove and in the vegetable garden. Upon request, she also produces filled pasta, Sardinian sweets, and other typical products with the sheep milk they produce, selling them to friends and acquaintances in town.

Animal production is pursued by Tonino, his brother, and his nephew. A semi-intensive production system involves a flock of around 700 sheep, kept in more than 120 hectares of land that is used for cereal, legumes, and hay for animal use. The sheep graze in the open 2–3 hours per day, but the flock spends most of the day in the stable where they are fed a mix of grains, hay, and concentrates so that they get the right amount of protein and fibre. He explained that if you leave sheep alone, they will eat anything they find. Fibre provides energy, gives a sensation of satiety, and ensures that there is enough fat in the milk. But the daily ration must have a well-balanced ratio of fibre to protein to allow optimal rumen functioning during digestion. Too much fibre can result in digestive problems for the animals. As Tonino explained in great detail, working out feed rations is one of the most important daily activities in the farm; they determine animals' welfare and, most importantly, their productive efficiency.

Everything is analysed, decided, and planned with the support of experts; nothing is left to chance. The impact that nature can have on the management of the farm is reduced to the minimum through in-barn feeding, preventive medicine, and monthly inspections by nutrition and veterinary experts. A commercial salesman (working for the Italian feed company from the Cargill group) supports planning animal nutrition and achieving the most productive and cost-efficient outcome so that milk production is maximized. A private veterinarian follows the reproductive

Figure 5.2 Tonino with his Unifeed wagon. It simplifies distribution of feed, reduces labour costs, and allows for controlled and standardized animal feeding. Villamassargia, south-west Sardinia, November 2020.
Credit: Giulia Simula.

cycle of the animals and helps Tonino with genetic selection and breeding of the animals, where choices are based on their productive capacity. Capital investment in the farm is very high and there is a strong reliance on external inputs as well. Without external feed, concentrates, and preventive medical care, the farm could not continue to exist.

Producing in a capital- and input-intensive way, and having a livelihood that mostly depends on the sale of milk and partly on the sale of lambs, means that when the price of milk goes down, Tonino has little or no flexibility to change the way of production or to diversify his income. According to Tonino, uncertainty can be controlled or reduced through careful planning, by following this cost-control and profit-maximization rationale that helps him to minimize external influences on the output produced. However, he is also highly unsatisfied because keeping everything under control is very labour- and capital-intensive. Tonino is now on a technological treadmill that feels like a trap from which it is hard to escape.

Although producers who first adopt certain technologies or practices to increase production might initially benefit financially, once other producers also adopt those technologies and increase their yields, the overall higher production results in the fall of commodity prices, and so the consequence is that producers must constantly increase their production to maintain the same revenue. The treadmill trap is quite evident when talking to Tonino. He works long hours without holidays and he still feels like all his sacrifices are not met with adequate returns. He is constantly menaced by uncertainty in the price of milk, which is exacerbated by trade policies that allow cheap meat and milk to come from abroad and be sold in local markets. Speaking in December 2019, he said:

> It's little, little, little. You work long hours and ... it's not paid ... If they paid me all the hours of work I do ... I get up at 4:00 a.m., come home when it's dark, what the f*** for? So that they collect all the money? No! You go protest, and for what? You don't achieve anything protesting.

So, even if Tonino paints a narrative where everything is under control and he and his brother can minimize uncertainties through investments and planning, he cannot control the price of milk and the price of inputs and so, despite all his investments, this does not result in higher revenues.

Since reducing inputs to production is not really an option, Tonino tries to manage uncertainty by building positive relationships with the industry to whom he sells milk. He has been selling to the same dairy industry for decades and, since he sells big quantities, he has some negotiating power when it comes to deciding the price. Even during periods of milk price crashes, he can secure a relatively constant milk price thanks to the personal relations he has built with the owner. This relationship is maintained through favours and gifts, and recently, he has been trying to convince a group of pastoralists to produce milk in summer so that the industrialist can produce all year round, resulting in an increase in revenue for Tonino as well.

While he intensifies production on one side, Tonino also has a long-term project. His son Piergiuseppe is 20 years old, and he is studying agronomy at the University of Sassari. Together with him, Tonino wants to de-intensify slowly and shorten the supply chain by producing cheese and selling directly to consumers. As he said:

> The world is changing ... I do not want to be in the hands of the industries anymore. Slowly my son Piergiuseppe will help me too, he's young now, but he'll come back with a degree. I'll put [in] the experience, he'll put [in] the theory, the knowledge.

To conclude, Tonino has always followed a productivist trajectory, based on the assumption that the intensification and rationalization of production with increased efficiency can lead to the control and prevention of uncertainty and to income increases. However, the reality is that the intensification of production leaves little flexibility to face uncertainty and results in high dependency on external inputs and on experts' advice and intervention. In this scenario, Tonino relies on personal relations with the industry to secure a decent price for his products, and at the same time he is trying to get out of the technological treadmill trap by building an alternative and gaining more control over his products.

Managing uncertainty by 'living in between': the flexibility of small-scale pastoralists and direct sales

Pastoralists with a smaller flock can face market uncertainties in a more flexible way. They sell to private or cooperative industries when it is convenient to do so, and they produce cheese and sell it directly to local communities and nested markets when the price of milk goes down.

Low-input production and co-production with community members is important to reduce costs. Such pastoralists also cut down on costs by controlling and adjusting the level of farm inputs. Smaller pastoralists also diversify income and make the most of the resources available in their territories – wood, forestry resources, olives, honey, and so forth. They tend to focus on other employment, especially when the milk price is low.

Relationships and social networks help them navigate uncertainty because such linkages help new forms of diversification to emerge. Social relationships are a source of help and reciprocity and so often provide a safety net (see Chapter 6). Women are key to maintaining and nurturing these social relationships; they attend social events and keep connected to the families in town. While older pastoralists are more connected to family and town-linked networks, younger pastoralists build relationships with urban communities and so broaden out their activities, with the involvement of the whole family. This might be through tourism and social activities, such as activities with schoolchildren, or opening the farm to young people with disabilities or to people who have to re-enter the labour market after completing a period of imprisonment. In a pastoral family, women are key to enabling diversified production and widening farm-based activities.

Felice, for example, has around 100 sheep. He milks by hand as he doesn't have a milking machine, and he is now building a small barn for the animals to avoid milking in the rain. To make this relatively small investment, he is waiting for the subsidy allocated to new, younger farmers, as he does not have capital available. Together with his family, he lives close to the northern coast of Sardinia, which is a tourist area and 10 minutes' ride from Sassari, one of Sardinia's major towns, in an area where tourists transit. He sells to a dairy cooperative located away from the coast, but when the price of milk is too low, he makes cheese and sells it to the local communities, tourists, and through local bars. His wife Elisa used to help Felice with the making and selling of cheese locally but now that they have three young children, she is fully engaged in care work. She prepares food, takes care of the house, and helps with the daily work that needs to be done in the house and the farm. Producing a small quantity of milk gives Felice space and flexibility to decide the destination of the milk according to the best conditions and to navigate uncertainty as it comes. In fact, on a 50-litre output, the difference between selling to the cooperative and making cheese at home is large. As Felice says:

> Last year it [one litre of milk] was 60 cents [selling to the cooperative]. Two years ago, it was 80. Three years ago it was 85. We are more or less around that figure. I am telling you again, on a 50-litre output, €35 to €37, it does not change my life. If you produce a thousand litres, from €700 to €750, it can make a difference because maybe with that €50 is earned daily, you pay a worker for example ... If I make cheese myself, I can get out 10 types of cheese and three ricotta ..., I can make €130. From one day's worth of milk, so now we're talking ... It's like selling for €3 a litre more or less. [As larger sizes were not selling], I've made smaller rounds of cheese. Once they are half mature, they weigh 800/850 grammes. I sell them for €10 a piece so in the end it is like selling the cheese at €12 a kilo. This way, I can sell it more easily because I can find €10 in the pocket of almost anyone.

Navigating market uncertainty also means being able to control farm costs. Felice's production is low cost and low input. This is both because he does not have the capital available and because he wants to rebuild those community relations of cooperation, reciprocity, and co-dependency that were the norm when his father was a pastoralist and which have now disappeared. He makes agreements with his neighbours who cultivate fruit trees to go and clean their field with his sheep after the harvest and take advantage of the stubble. During the olive harvesting period, he also goes to the oil mill to get the olive leaf waste. He then spreads the freshly cut leaves on his two hectares of land as fodder for the animals. The leaves are also useful for fertilizing the land. He also ploughs his neighbour's field before sowing and, in return, he uses his neighbour's cooler to store milk before the cooperative's truck picks it up.

These social dynamics were everyday practice for a pastoralist until the 1980s and 1990s but with policy incentives focused on intensifying land and increasing the use of technology, flock sizes increased and so did milk production. Pastoralists then started working more individually using family labour rather than reciprocally helping each other in groups. However, during periods of repeated crisis and increased uncertainty, these practices remain key to navigating market volatility and shrinking revenues, and pastoralists like Felice are investing in them once again.

Income diversification is important for reducing farm costs and boosting income. Working in the farms of bigger producers and landowners serves both to build a network of collaboration and reciprocity and to diversify one's income. Small-scale pastoralists have some flexibility and, although a small flock living on reduced inputs does not produce as much milk and as much income, it can survive and needs less constant care. In this way, the pastoralist families can free up some time for seasonal work and for trading in homemade products.

Figure 5.3 Felice leading the flock to graze in the land he accesses through informal agreements with his neighbour. Sorso, north-west Sardinia, January 2020.
Credit: Giulia Simula.

Reducing farm inputs is also vital. While there is an awareness that many farm inputs are industrial products – for example, feed comes from soya or other genetically-modified grains grown on a large scale in eastern Europe, Asia, and Latin America – small-scale pastoralists must take a pragmatic approach. They might prefer to use other types of feed, but they also know that some supplementation is necessary to have a production that can sustain the family income. As a result, they increase feed supplements when the price of milk rises and reduce their use by favouring open grazing when the price of milk falls. This is not experienced as a contradiction but as a necessary coexistence with an uncertain context. The same goes for medical care. Medical intervention is often kept to the minimum and left to times when strictly necessary. Animals are not squeezed as productive machines but rather pastoralists try to ensure that the animals become accustomed to coping with uncertain circumstances as part of a wider approach to pastoral care.

For those who sell artisanal cheese through their own shops, engagement with a more formal market system adds costs. In order to meet sanitary standards, investments in processing and selling facilities – improving buildings, investing in refrigeration, and so on – are imperative. As soon as you enter the formal market, taxes must be paid too, book-keeping and accounting must be in order, and all regulations must be followed. This is costly compared to the informal systems of production and sale. This in turn means that costs of production have to be kept in check and the marketing of products improved if the whole chain is to be profitable. Although Felice does not have his own shop, market networks remain essential.

Across more formal and more informal production and marketing, family and territorial networks are central in times of uncertainty. Selling in niche markets (where quality, territoriality, and the social, cultural, and traditional value of the product are the focus) requires much skill and good collaborative networks. While among those who sell formally these agreements may take the shape of a written contract, among informal producers, they are just verbal agreements. Even formalized mini-dairies, however, must sustain themselves through the extensive networks they build based on shared values between producers and consumers. Moreover, this flexibility to sell milk to the industry or to turn it into cheese for sale allows producers to respond to uncertain market conditions as they unfold.

Conclusion

For decades, pastoralists have been dismissed as 'backward', ignorant, and unwilling to develop and modernize. This modernizing narrative has contributed to transform pastoralism in Sardinia. But this narrative is based on the assumption that everything can be planned, risks can be calculated and controlled, and market fluctuations can be overcome by increasing efficiency and investing in technology. Yet pastoralists' logics, rationalities, and practices start from the notion that conditions are variable and cannot always be controlled. They are experts in dealing with uncertainty and they have much to teach us in this regard.

Flexibility, especially when dealing with market uncertainty, is key. This means that being able to remain in the informal sector is very important for small-scale producers in order to avoid high costs of production and the productivity trap. The value of pastoralism goes beyond the production of particular commodified outputs. Pastoralists also invest in and build communities and social networks. They also have an important environmental function, managing pastures and supporting biodiversity. They are also innovators, creating new agri-tourism activities and also linking with urban communities.

As a result, policymakers should be aware of the wide variety of pastoral systems and their numerous positive effects on society and the environment. Instead of applying top-down models of development, the contextual richness and variety of pastoralism in practice must be acknowledged. While controlling uncertainty through investments in technology aimed at boosting production efficiencies may work for some, there are challenges, as the case of Tonino showed. Meanwhile, small-scale pastoralists, such as Felice – who are by far the majority – take a more flexible approach. They are the real experts at living with and responding to uncertainty. Policies and public services need to take such experiences into account, and indeed learn from them.

References

Atzori, A.S. et al. (no date) *Report of the Characterization of Sardinian Dairy Sheep Production Systems*. Available from: <http://www.sheeptoship.eu>.

Farinella, D. (2018) 'La pastorizia sarda di fronte al mercato globale. Ristrutturazione della filiera lattiero-casearia e strategie di ancoraggio al locale', *Meridiana: Rivista di Storia e Scienze Sociali* 93: 113–134.

Farinella, D. and Simula, G. (2021) 'Ovejas, tierra y mercado: dependencia de los mercados internacionales y cambios en la relación entre pastores y naturaleza', *Relaciones Internacionales* 47: 101–24 <https://doi.org/10.15366/relacionesinternacionales2021.47.005>.

Ismea (2019) *Il Mercato dei Formaggi Pecorini: Scenario Attuale e Potenzialità di Sviluppo tra Tradizione e Modernità dei Consumi*, Ismea, Roma.

Mencini, G. (2021) *Pascoli di Carta: Le Mani sulla Montagna*, Kellermann Editore, Vittorio Veneto.

Sardegna Agricoltura (2013) *Il 6° Censimento Generale dell'Agricoltura in Sardegna: Caratteristiche Strutturali delle Aziende Agricole Regionali*, RAS, Cagliari. Available from: <http://www.sardegnastatistiche.it/documenti/12_103_20130710170153.pdf>.

Simula, G. (2019) 'Should we cry over spilled milk? The case of Sardinia', in PASTRES [blog] <https://pastres.org/2019/02/15/should-we-cry-over-spilled-milk-the-case-of-sardinia/> (posted 15 February 2019).

Simula, G. (2022) 'Pastoralism 100 Ways: Navigating Different Market Arrangements in Sardinia', doctoral dissertation, Institute of Development Studies, University of Sussex, Brighton. <https://sro.sussex.ac.uk/id/eprint/109485/>

CHAPTER 6
Responding to uncertainties in pastoral northern Kenya: the role of moral economies

Tahira Mohamed

Introduction

Pastoralism is a vital source of livelihood for about 10 million people across northern Kenya. Yet events such as droughts, floods, disease outbreaks, locust attacks, conflict, and displacement affect pastoralists in this region. Over the past decades, the frequency and intensity of such challenges in the drylands of northern Kenya have increased due to climate change, shifts in land tenure security, and transformations in political economy (Birch and Grahn, 2007; Mohamed, 2022). In pastoral areas, such events may occur in sequence or may compound many already-existing challenges. They cannot be predicted and so events remain fundamentally uncertain.

However, by failing to embrace uncertainty, standard development approaches often fail. They are instead premised on predicting, managing, and controlling pastoral systems. These approaches include disaster risk management through early warning systems, shock-responsive social protection, and market infrastructure development (Bailey et al., 1999; Borton et al., 2001; Barrientos et al., 2005; Caravani et al., 2021; see also Chapter 1). Many standard approaches see the route to controlling risks as through standard investments in productivity and market development (Catley and Aklilu, 2012; Carter et al., 2018).

For example, in northern Kenya, between the 1970s and 1980s, the focus was on food aid and irrigated agricultural expansion to support food security. Rangeland development emphasized water infrastructure, livestock feed supply, and reseeding programmes, among other interventions. In the 2000s, Kenya's National Drought Management Authority emerged as the leading agency managing drought and related disasters through providing early warning bulletins and contingency planning. Later, numerous pastoral livelihoods and resilience-building projects were developed to promote livestock market infrastructure, abattoirs, and investment in the livestock value chain. Such production-oriented interventions were combined with social protection through cash transfers, livestock insurance, and the promotion of 'climate-smart' agriculture to replace traditional food security projects (Mohamed, 2022).

Although there have been some successes in responding to disasters in pastoral areas via project-based interventions, such as supporting livelihood diversification and initiating cash transfers (Little, 2001), pastoralists are still affected by climate-related uncertainties, conflict, and social instability. Over decades, pastoralists have thus become enlisted in massive aid programmes, resulting in self-reinforcing dependency, with pastoralists seen as the victims of recurrent disasters.

Given this experience, this chapter asks whether the state, humanitarian agencies, and development interventions have missed their mark by focusing on predicting and controlling risks rather than embracing and managing uncertainties as part of continuous, everyday practices of generating reliability. Could pastoralists themselves, through their adaptive strategies and redistributive moral economy practices, show us an alternative approach more attuned to dryland uncertainties?

The chapter argues that pastoralists should not be seen as passive victims of disaster, forever reliant on external support, but that they have their own agency; their own practices embedded in social relations that help them respond to complex, uncertain, and unpredictable events. Living with and from uncertainty is central to pastoral livelihoods, and it should be fundamental to the disaster response policies and development strategies in pastoral areas.

Responding to uncertainties: the role of pastoralist moral economies

Pastoralists respond to uncertain events by galvanizing various relationships embedded in social-cultural and religious identities. These relationships allow for the redistribution of resources, including livestock and labour, as well as extending solidarities in times of need through moral economies. Moral economy practices based on collective redistribution of resources and solidarities are essential to how pastoralists manage uncertainties.

The concept of 'moral economy' gained popularity following the classic 1971 essay by E.P. Thompson, *The Moral Economy of the English Crowd*, and James Scott's *Moral Economy of the Peasant* in 1976. Both Thompson and Scott used the concept of moral economy to explain collective resistance against food crises, exploitation, and unfair practices. However, Scott also emphasized the notion of 'subsistence ethics' that facilitated reciprocity and redistribution. In a pastoral context, a moral economy enshrines reciprocity, redistribution, social insurance, and the formation of identities that are essential in helping people survive and thrive, including under uncertain conditions (Ensminger, 1992; Bollig, 1998; Mohamed, 2022).

Many standard disaster risk reduction and social protection policies in pastoral areas ignore these vital moral economy relationships between diverse social groups, including men and women, the wealthy and poor, and young and old. Through case studies in two pastoral settings in Isiolo County, one urban and another remote, this chapter reveals the fundamental role of such culturally and religiously embedded relationships that help pastoralists respond to uncertainties.

The study sites

The arid and semi-arid area of Isiolo County is about 250 km from Kenya's capital, Nairobi (Figure 6.1). Isiolo borders five other pastoral counties, including Marsabit, Samburu, Garissa, Laikipia, and Wajir. The county also borders the farming areas of Meru county as well as Meru National Park, creating competing land uses. The area is traversed by the Ewaso-Nyiro river that is a significant resource for humans, wildlife, and livestock. In the early 1970s, Isiolo Town had a human population of about 21,000, poor infrastructure, and livestock-keeping as the main livelihood of the population (Republic of Kenya, 1976). From the 1970s to date, Isiolo has grown into a modern town with a high population, diversified livelihoods, and significant infrastructure, including an international airport, major abattoir, government institutions, and international NGO offices. As of 2019, the human population stood at 268,000 (KNBS, 2019). Although the rise in population and growth of the urban centres is clearly visible, the remote regions of the county remain under-developed, with poor roads and services.

Fieldwork for this chapter took place between 2019 and 2022 in two pastoral areas – Korbesa and Kinna – the former remote and rural and the latter more urban, and also closely linked to the major town of Isiolo and nearby Meru (Figure 6.1). Through a series of community and historical event-mapping exercises, key informant interviews, focus group discussions, in-depth narrative interviews, and photovoice exercises, I explored how pastoralists in Korbesa and Kinna managed uncertainties through time and space.

Korbesa is located in the very arid northern region of Merti sub-county, about 290 km from Isiolo town, and receives an annual rainfall of about 200 mm (Republic of Kenya, 2018). Korbesa people are predominantly pastoralists keeping cattle, camels, and goats. The owners herd the livestock through partnerships with relatives and friends and with support from hired labourers. Korbesa is a small centre and depends on nearby Merti town for services, including water, a market, and transport to other parts of Isiolo. Due to increased insecurity and livestock-raiding, pastoralists practise livestock redistribution to enhance their livelihoods. However, as the case studies below show, these redistributions are uneven, stratified between wealthy and poor people.

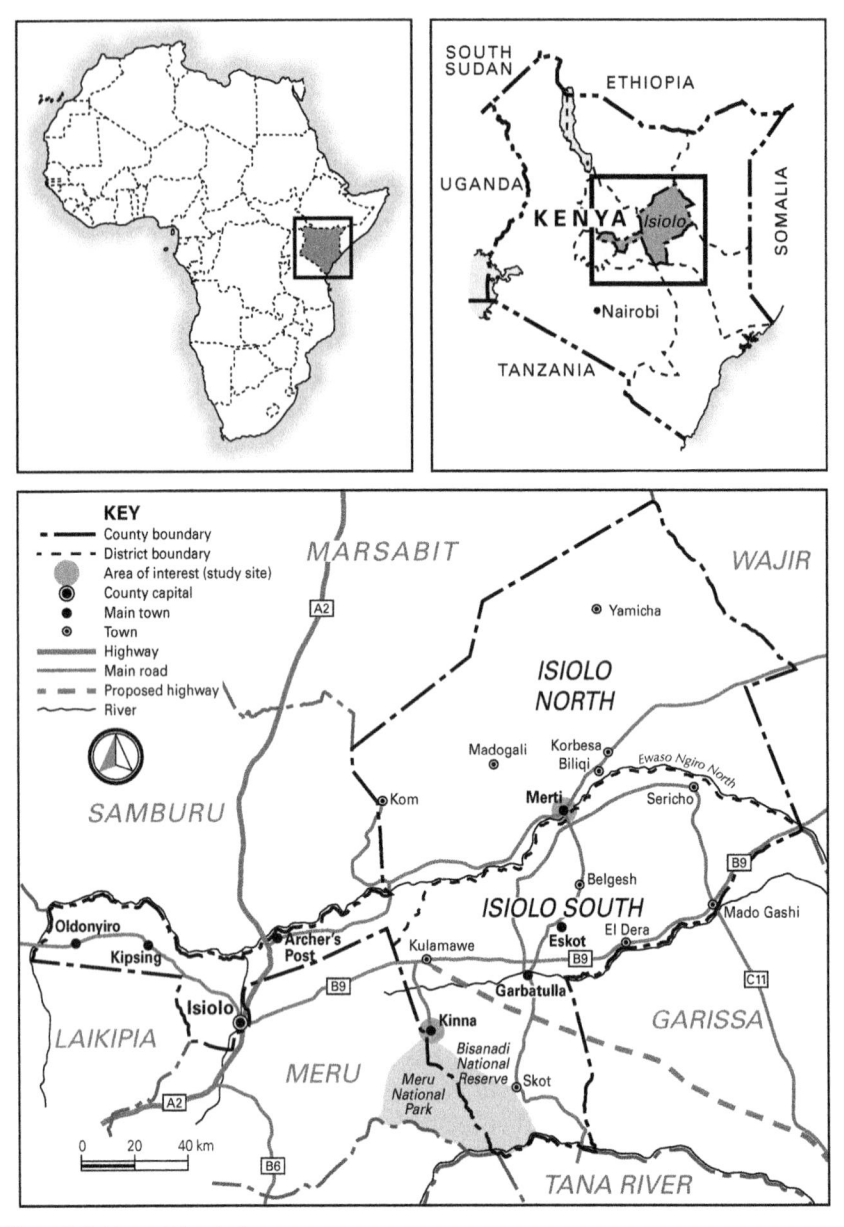

Figure 6.1 Map of the study area

Kinna is in the southern, semi-arid region of Garbatula sub-county, about 90 km from Isiolo town, and receives an average annual rainfall of 500–600 mm (Republic of Kenya, 2018). The inhabitants of Kinna practise livestock-rearing in combination with some small-scale irrigated farming, while others have diversified into trade, mining, and motorcycle transport services due to good connectivity to major towns. Insecurity and livestock-raiding are common due to the competition over resources, fuelled by the increased presence of small firearms and weapons spread through porous borders and armed conservation rangers. Another significant challenge is tsetse fly infestation through elephant-inhabited habitats, especially near the rivers.

Recurrent droughts and prolonged dry periods wreak havoc regularly in Kenya's pastoral regions, including Isiolo. This has been especially acute in recent years, with international humanitarian organizations reporting rainfall failure, livestock deaths, and the subsequent loss of livelihoods and acute hunger in the area during 2022 (United Nations, 2022). The Kenya Food Security Steering Group estimated that close to 4.4 million people in Kenya's arid and semi-arid counties were subject to a severe food crisis due to the failed rains over three consecutive seasons (Republic of Kenya, 2022). Emergency appeals have raised the alarm, and pleas for humanitarian aid to cushion the pastoralists from the increasing catastrophe have been made.

Despite the numerous development projects in these areas promoting food security, climate-smart agriculture, rangeland development, market intensification, and resilience building, pastoral areas continue to experience drought-related emergencies. Clearly, such development efforts have not worked. What lessons can we learn from pastoralists' own practices, and in particular how they draw on different relationships and moral economy practices to respond to and manage various uncertainties?

Pastoralists' strategies in responding to uncertainties in northern Kenya

Moral economy practices enhance resource redistribution and foster collective solidarities and comradeship to help manage uncertainties, including those due to drought, animal disease, livestock-raiding, and labour deficits. The bonds of families, religious networks, neighbourliness, and social-economic ties remain instrumental in facilitating solidarities and redistribution via such moral economy relations. The following four cases illustrate this dynamic.

Case 1: Uncertainties are everyday business

In February 2020, I met Salado, a 70-year-old female herder, grazing her sheep and goats in Bibi, a grazing camp, about 16 km from Kinna town in Isiolo County. Salado was constructing small pens for her goats' kids, and the thorny shrubs pricked her hands when she fed the young goats. We sat down and began our conversation, and Salado started to explain how multiple uncertainties surround animal production. She said:

> The frequency of *ola* (absence of rain) has increased, and in 2017, we lost 50 goats and were left with 50 goats. The remaining goats are weak because of the failure of the 2019 rain. Now we are afraid as the *ganna* (April rain) might fail. Before the animals recover from the preceding droughts, disease and subsequent drought affect

> reproduction. Since 2017, my goat has not reproduced, and multiple conditions threaten its growth. There are rising cases of *gandhi* (trypanosomiasis), and I have lost 18 goats. Sometimes, the disease outbreak coincides with a severe dry period making life difficult. In this place [Bibi], there are a lot of hyenas and if the sun sets while the livestock is not back in the pens, then expect nothing. Both the sunset and the hyenas compete. We must be actively involved in scouting pasture, tending to young and sick animals, and staying vigilant to guard against enemies, wild animals, and livestock raiders. Since I could not manage all these, I sought help from a relative, who gave me a herd boy.

Salado describes how she lives every day in fear, requiring persistent alertness. Her case uncovers the significance of knowledge about and preparedness for multiple uncertain conditions that arise in livestock production and reduce the productivity of the animals.

Case 2: Redistribution via clan and mosque

Asha is an elderly widow who lives in Biliqi village on the outskirts of Korbesa sub-location in Merti sub-county. Her husband Dido passed on 30 years ago. She currently lives with her elderly mother, her co-wife Raheema, her son Boru, and five grandchildren. In an interview in the Eman Hotel in Isiolo, she explained:

> In the 1996 *ola istaaga* (the standing drought), we lost six cattle and were left with four. My clan member, Barasa, invited our family to move from Korbesa and live in his neighbourhood, Biliqi. Barasa then took custody of our livestock and herded them for free so that I could travel to Isiolo town to seek help from family and friends. During the 2017 drought, the Samburu raided livestock from Borana, and we lost 30 cattle and lived on only six milk cows that survived the raid. To help recovery, I received one heifer from my clan member, and my co-wife, Raheema, as well as receiving a *dabare* (loaned) animal from a relative. Three months after the raid, during the Islamic month of *Muharram* [the first month of the Hijri calendar], the mosque organized a collective *zakat* [Islamic tax on wealth] and distributed animals to vulnerable people and victims of the raids. We received four goats to sustain our family. The mosque also gave three cattle to orphans' families left behind by the victims of the attack.

Although suffering major losses through livestock-raiding, Asha was able to recover due to support from her clan members, as well as the mosque. Asha's case reveals the importance of lineage relationships, mosque collection, and redistribution in enhancing social support, especially for lower-income families and widows.

Case 3: Comradeship and resource pooling

I met Haro, a young male herd-owner, in Merti sub-county in Lakole while conducting a focus group discussion in January 2020. Haro requested a ride to Merti to visit his family after spending weeks in Dogogicha, a borehole in the drought reserve, about 40 km from Merti. While we were travelling, we talked casually, and he narrated how he had been managing his herd in the past droughts:

> It has not been easy living through a series of dry periods, intensified insecurities, and livestock burglary from Kore [Samburu]. In 2013, I banded with three of my friends [Molu, Guracha, and Adi] Each of us hired an extra herder and contributed money to obtain a firearm to guard the animals. In the 2017 drought, we jointly purchased a motorcycle to facilitate movement and pasture surveillance in the rangeland. In the recent *adolessa* (prolonged dry season), we moved to a place called Qori, dug shallow wells, and settled with our livestock. When we depleted the pasture, we purchased feed from Isiolo and managed our animals. We also stored sufficient *dibu* (pesticides) and veterinary supplies to control disease spread. Luckily, we only lost four *yabiyye* (calves). In a good season, the four of us manage the animals in shifts, so we get extra time to spend with families in town. However, in a difficult time, we all work together by providing specialized attention, digging wells, and enhancing security. Working together helped us access *waheel* (comrades) and resources to manage our animals.

Haro explained how he combined labour resources and technology to survive multiple dry seasons, and that personal friendships allowed for the pooling of resources. Young, less asset-rich herd-owners cobble together resources to manage the increasingly commodified livestock practices. They make use of modern technologies, including motorcycles and firearms, to protect the livestock and enhance safe movement. Working with others allows friends to confront uncertainty together.

Case 4: Diversification and gendered networks

Bisharo is a member of a women's group in Isiolo town that specializes in the production and processing of camel meat called *nyirinyiri* in Somali or *koche* in the Borana language. In El Boran Hotel in Isiolo town, I had an in-depth livelihood interview with Bisharo, who narrated her livelihood trajectory:

> I began working as a retail trader bringing clothes and other items from Nairobi. In 2017, my friend Asha introduced me to the *nyirinyiri* group comprising about 15 women who process about 80–100 kgs of camel meat daily and export it to Nairobi. The members meet daily at the chairlady's house in Isiolo town to process the meat. We buy camel meat from butchers in Isiolo town, or sometimes we purchase live animals

and ask the butcheries to slaughter them on our behalf. After slicing and sun-drying the meat, we cut it into small chunks and oil-fry it. We send the meat to Nairobi's Eastleigh market to retailers, hotels, and events such as weddings and parties. To reduce transport costs, we use a lorry that belongs to the Anolei camel milk women's group [a group of women's enterprises that export camel milk from Isiolo to Nairobi]. Some members of the *nyirinyiri* group are also part of the Anolei group; therefore, the connection helps us market our products and get customers, especially for weddings and other events. Every Sunday, we have a weekly meeting to assess our progress and raise funds to support our members. Every member contributes KES 500 [US$4] weekly; whenever a group member faces shortages or the loss of a family member, we contribute from our savings. We provide loans if a member needs a loan to pay school fees or for other needs. This group has enhanced our capacity as women, and we have continued income for food and other necessities. We aim to expand our production, build a modern kitchen and working space, and become international camel meat exporters.

Bisharo's case highlights the importance of trust and networks for connecting to diverse markets, and so ensuring that a diversified livelihood linked to pastoralism can address uncertainties. Her case shows the significance of the gendered network between women in camel meat and milk production and the importance of savings to enhance social support systems for life-cycle events such as deaths, weddings, and illness.

Together, the case studies reveal how cascading uncertainties and variable conditions such as drought, animal disease, and insecurities from wild animals and livestock-raiding affect pastoral livelihoods. Different pastoralists manage uncertain conditions in a variety of ways but always drawing on social relations. The form these responses take depends on wealth, kin connections, religious affiliation and commitments, access to technologies, and diversification opportunities. Central to responding to uncertainty is investing in relationships and networks through *milo* (lineage) and *hariyya* (friends). Equally, the pooling of resources as part of collective responses is essential to community solidarities and helps pastoralists respond to diverse, uncertain conditions.

How then do these approaches of collective mobilization of social and economic resources, redistribution, comradeship, and solidarity differ from the risk-focused approaches of standard external interventions such as social protection, livestock insurance, and disaster response?

Pastoral practices and external interventions

Building resilience amidst the intersecting uncertainties discussed in this chapter is a crucial development agenda. But are current development approaches the answer? As already discussed, most standard interventions focus on risk (where futures can be calculated, predicted, and therefore anticipated and planned for), whereas pastoralists must respond to uncertainties (where futures are not known, either in terms of likelihoods or outcomes) (see Chapters 1 and 9). Risk-oriented approaches frequently envision an individualized response to a defined risk (a drought, a disease, etc.), whereby targeted interventions (an insurance product, a cash transfer, a livelihood project) will return people's livelihoods to a stable norm.

Instead, as the case studies show, responding to uncertainty must be a continuous affair, with intersecting and cascading uncertainties dealt with together and as part of everyday practices in a highly variable environment. Multiple responses, suited to different contexts, diverse shocks, and contrasting personal situations must be combined. So, for example, people may call on relatives or the mosque, and frequently work together as a group in mutual solidarity. Such moral economy practices are rooted in cultural identities and embedded in social relations and are not simple 'interventions' as part of a disaster risk response or social protection programme. The following four themes emerge from this analysis.

Actively embracing uncertainty

Contrary to the costly humanitarian aid and development projects premised on early warning systems, anticipation, monitoring, and controlled planning meant for stable settings, pastoralists live in an unpredictable environment and must learn to adapt through innovative practices. As shown across the cases, pastoralists embrace a sense of unfolding, continuous time, expecting surprises and unforeseen events that could threaten a potentially good season for their animals. They must always be prepared, ready for the unexpected. This is normal life. Pastoralists therefore must always stay alert and informed, always engaging in innovative learning and timely responses.

As Salado noted, the coming rainy season might be accompanied by animal disease outbreaks or flooding, affecting livestock productivity. There is therefore a need to embrace uncertainty as a continuous flow of events and not a one-time shock, represented by a specific risk or disaster. Conventional risk assessment and the management of project-based interventions, such as targeted social protection and livestock insurance, miss out on the fluidity of extended, unfolding time and of contingent uncertainty, and instead measure risks in terms of exposure to identifiable and predictable events.

For pastoralists' resilience in the face of unknown events to be enhanced, it is vital for development policies and humanitarian support to shift their focus. There is a need to acknowledge uncertainty and how this is actively responded to by 'reliability professionals' in pastoral systems (Roe, 2020). Reliability professionals must act in real time through learning and sharing information, and so necessarily embrace uncertainty, managing responses while avoiding the dangers of ignorance. Development projects and humanitarian support could expand their scope and incorporate such local perspectives, instead of imposing standard models. These models assume stability,

Figure 6.2 '*Borantiti* [being Borana and showing the ideals of the Borana] is all about showing kindness and solidarity to overcome shortages. Here, women share labour to load water on the donkey.'
Credit: Bushra (from the photovoice exercise).

singular risks, and that resilience can be created through a productivist model of livestock development in contexts where uncertainty and unpredictability are central to people's livelihoods.

Identities, relationships, and collaboration

Investing in multiple relationships based on lineage, friendship, religious networks, and economic ties through moral economies is essential, whether through redistribution after losses or providing a coping mechanism during a crisis, as the cases show. As one Borana saying goes, *Hoorin dumansaa naamum arra qabuut bor daaba* (Livestock are like clouds, they can precipitate at any moment, and those we have today might be lost tomorrow), hence the importance of sharing. Identities and relationships are paramount for pastoralists who must collaborate to generate reliable support in difficult times. As shown across all cases, different relations and collective solidarities have been vital in providing access to resources, offering security in the rangelands as well as market connections.

By contrast, many development projects aim to support 'vulnerable' individuals, dissociated from their social and cultural contexts, through predetermined metrics used to target distribution in social safety protection programmes, for example. Too often, such programmes, whether around cash transfers or insurance programmes, ignore the collective social and cultural practices essential for managing uncertainties. For instance, Kenya's Livestock Insurance Programme aims to provide insurance cover to pre-selected households based on a certain vulnerability threshold. These projects however fail to acknowledge the dynamic relationships central to livestock sharing and redistribution, centred on social, cultural, and religious practices (see Chapter 7). But, if project-based livestock transfers are combined with forms of collective redistribution, such as *zakat* through mosques, reliable livelihoods and collective ownership could be enhanced.

Adaptive technology

Although pastoralism is perceived by some as a 'backward' livelihood and resistant to change, the cases have shown how pastoralists adaptively use modern transport and communication technologies to respond to uncertainties. Even in the marginal areas far from town, pastoralists collaborate and pool resources to acquire motorcycles to enhance mobility and market connections. The spread of mobile phone communication in pastoral areas has improved efficiency in sending money and transacting trade. This shows how pastoral livelihoods are flexible and adaptable to changing political economy and development, making productive use of technology. Although access to technology is unequally distributed, pastoralists pool resources and overcome limitations.

In the same way, development projects and humanitarian support must incorporate accessible technologies in their programmes to enhance efficient service delivery in responding to uncertainties in pastoral areas. For example, mobile phone technologies are vital in disease surveillance in pastoral areas, as well as in responding to variable fodder and water sources (Ikiror et al., 2020).

Networking, trust, and diversification

As the case of the *nyirinyiri* group showed, informal groups connect many people in pastoral areas and have multiple functions. Although focused on processing and selling camel meat, the group is also engaged in savings, as well as being linked to camel traders in the market, butchers in town, female processors of camel milk and meat, transport providers, and final retailers in Nairobi. Such networking, central to any market engagement that can respond to uncertainty (see Chapter 5), is enhanced by the availability of transport and mobile money transfer technologies, facilitating transactions, communication, and movement of goods.

Networks are nurtured by trust and the web of relationships between the parties involved. The result is a reliable flow of goods with limited interruption, even in the face of considerable uncertainties. For example, during the Covid-19 pandemic, despite movement restrictions and market closures, the *nyirnyiri* group continued producing meat for their market through their diversified networks. Aside from the market relationships, the members' saving club provides women with continued social support and emergency loans to manage shortages.

By contrast, standard development and humanitarian interventions are often focused on single projects, which have limited flexibility and fail to develop trust during implementation, despite the rhetoric about bottom-up and 'participatory' approaches. Styles of implementation, forms of audit control, and strictly defined protocols and plans in aid bureaucracies can therefore undermine the capacities to respond to uncertain conditions (Caravani et al., 2021).

Conclusion

This chapter has shown how pastoral production in northern Kenya is full of uncertainties and unpredictable surprises that arise due to climate-induced events such as drought or floods, animal disease, and political and economic challenges that undermine pastoralists' livelihoods. Pastoralists respond and adapt to these uncertainties with multiple practices embedded in social, cultural, and economic ties, centred on what has been described as pastoral moral economies.

However, such responses have their limits. First, the responses are highly stratified and differentiated, especially between the wealthy, well-connected herd-owners and low-income families. While redistributive and collective measures help the poor and marginalized, richer pastoralists remain in a better position than the most vulnerable members of society. It is therefore crucial to acknowledge the differentiated capacity of diverse social groups, including the wealthy/poor, women/men, young/old, and pastoralists in towns/remote areas. Second, when shocks overwhelm – as with recurrent and extensive drought or a widespread pandemic – local moral economy practices for sharing, redistribution, and mutual assistance may be insufficient. For this reason, state-led responses to disasters should complement local responses and help improve local capacities rather than displacing them.

In pastoral areas with limited infrastructure and weak state capacity, especially in settings that are insecure, external support is in any case patchy. Here, supporting local practices and local 'professionals' able to generate reliability is essential if resilience is to be generated. Parachuting in projects that aim to generate 'smart' responses to complex challenges is inadequate. All responses need to be embedded in local networks, and build on trust and existing social relationships. Centralized, top-down designs do not work, even if they are aimed at humanitarian and developmental outcomes. Building the capacity of local communities to generate resilience through their own practices, supported by external intervention as appropriate, must accept that simple plans focused on smart intervention and anticipatory approaches are unlikely to work.

Instead, building resilience and responding to uncertainties in pastoral areas should build on and help establish relationships and trust entrenched in everyday moral economies, thereby enhancing reliable and flexible livelihood amidst surprises. Pastoral livelihoods will continue to thrive and adapt to turbulent and uncertain times if external intervention invests in these local collective solidarities, while assisting with the timely redistribution of resources as well as investing in networks and relationships. Through such routes, flexible and reliable adaptation to what is always a variable livelihood production system will be achieved.

References

Bailey, D., Barrett, C.B., Little, P.D. and Chabari, F. (1999) *Livestock Markets and Risk Management among East African Pastoralists: A Review and Research Agenda*, SSRN Scholarly Paper 258370 <http://dx.doi.org/10.2139/ssrn.258370>.

Barrientos, A., Hulme, D. and Shepherd, A. (2005) 'Can social protection tackle chronic poverty?' *European Journal of Development Research* 17: 8–23 <https://doi.org/10.1080/09578810500066456>.

Birch, I. and Grahn, R. (2007) *Pastoralism: Managing Multiple Stressors and the Threat of Climate Variability and Change*, UNDP Human Development Report Office Occasional Paper, New York NY.

Bollig, M. (1998) 'Moral economy and self-interest: kinship, friendship, and exchange among the Pokot (N.W. Kenya)', in T. Schweizer and D.R. White (eds), *Kinship, Networks, and Exchange*, pp. 137–57, Cambridge University Press, Cambridge.

Borton, D., Morton, J. and Hendy, C. (2001) *Drought Contingency Planning for Pastoral Livelihoods*, Natural Resource Institute, Chatham.

Caravani, M., Lind, J., Sabates-Wheeler, R. and Scoones, I. (2021) 'Providing social assistance and humanitarian relief: the case for embracing uncertainty', *Development Policy Review* 40: e12613 <https://doi.org/10.1111/dpr.12613>.

Carter, M.R., Janzen, S.A. and Stoeffler, Q. (2018). 'Can insurance help manage climate risk and food insecurity? Evidence from the pastoral regions of East Africa', in G. Lipper, L. McCarthy, N. Zilberman, D. Asfaw and S. Branca (eds), *Climate Smart Agriculture: Building Resilience to Climate Change*, pp. 201–225, Springer International, Cham., Switzerland.

Catley, A. and Aklilu, Y. (2012) 'Moving up or moving out? Commercialization, growth and destitution in pastoralist areas', in A. Catley, J. Lind and I. Scoones (eds), *Pastoralism and Development in Africa: Dynamic Change at the Margins*, pp. 85–97, Routledge, London <https://doi.org/10.4324/9780203105979>.

Ensminger, J. (1992) *Making a Market: The Institutional Transformation of an African Society*, Cambridge University Press, Cambridge.

Ikiror, D. et al. (2020) 'Livestock disease surveillance through the use of smart phone application in Isiolo County Kenya', *East African Journal of Science, Technology and Innovation* 2: 1–16 <https://doi.org/10.37425/eajsti.v2i1.218>.

KNBS (2019) 'The Kenya population and housing census: population distribution by administrative unit', vol. 1A, Kenya National Bureau of Statistics, Nairobi. Available from: <https://www.knbs.or.ke/download/2019-kenya-population-and-housing-census-volume-ii-distribution-of-population-by-administrative-units/>

Little, P. (2001) *Income Diversification among East African Pastoralists*, Research Brief 01-08-PARIMA, Global Livestock Collaborative Research Support Program, University of California, Davis CA. Available from: <http://crsps.net/wp-content/downloads/Global%20Livestock/Inventoried%207.12/2-2001-2-116.pdf>

Mohamed, T. (2022) 'The Role of the Moral Economy in Response to Uncertainty among the Pastoralists in Northern Kenya', doctoral dissertation, University of Sussex <https://sro.sussex.ac.uk/id/eprint/110472/>.

Republic of Kenya (1976) 'Isiolo District Development Plan 1974–1978', Government Printer, Nairobi.

Republic of Kenya (2018) *Isiolo County Integrated Development Plan, CIDP, 2018–2022: Making Isiolo Great*, Republic of Kenya. Available from: <https://repository.kippra.or.ke/bitstream/handle/123456789/1409/2018-2022%20Isiolo%20County%20CIDP.pdf?sequence=1&isAllowed=y>.

Republic of Kenya (2022) *The 2022 Long Rains Season Assessment Report*, Kenya Food Security Steering Group (KFSSG), Republic of Kenya. Available from: <https://www.ndma.go.ke/index.php/resource-center/send/86-2022/6602-lra-national-report-2022>.

Roe, E. (2020) *A New Policy Narrative for Pastoralism? Pastoralists as Reliability Professionals and Pastoralist Systems as Infrastructure*, STEPS Working Paper 113, STEPS Centre, Brighton. Available from: <https://opendocs.ids.ac.uk/opendocs/bitstream/handle/20.500.12413/14978/STEPS_WP_113_Roe_FINAL.pdf?sequence=105&isAllowed=y>.

Scott, J.C. (1977) *The Moral Economy of the Peasant: Rebellion and Subsistence in Southeast Asia*, Yale University Press, New Haven CT.

Thompson, E.P. (1971) 'The moral economy of the English crowd in the eighteenth century', *Past & Present* 50: 76–136.

United Nations (2022) 'Severe drought threatens 13 million with hunger in Horn of Africa', *UN News*, 8 February 2022. Available from: <https://news.un.org/en/story/2022/02/1111472>.

CHAPTER 7

Livestock insurance in southern Ethiopia: calculating risks, responding to uncertainties

Masresha Taye

Introduction

Pastoralists in the Horn of Africa face a plethora of complex and intertwined risks and uncertainties (Homewood, 2008). These include climate change, rangeland fragmentation and privatization, conflict, and shifts in governance regimes (see, for example, Little et al., 2012; McPeak et al. 2012; Catley et al., 2013; Lind et al., 2020). In particular, recurrent climate-induced shocks, exemplified by frequent droughts, have been identified as a key challenge facing pastoralists (Carter et al., 2007; Barnett et al., 2008). As a result, many policies and programmes have been developed to address climate-induced vulnerabilities on pastoral families and communities.

One of the most recent disaster risk finance mechanisms responding to drought impacts in pastoral areas is Index-Based Livestock Insurance (IBLI). It has been lauded as an innovative, market-driven technique for managing drought risk for the most vulnerable pastoralists. Proponents claim that when drought strikes a region and payouts are distributed, insured pastoralists will respond to the drought risk by investing in feed, water, and veterinary services to keep insured animals alive. As a result, it is claimed that IBLI is a 'pro-poor' development intervention, focused on protecting key assets, rescuing pastoralists from further poverty traps (Barnett et al., 2008; Chantarat et al., 2009).

However, does making such a product available to all pastoralists on a commercial basis achieve the desired purpose of aiding the poorest and most vulnerable in a pastoral system? Does focusing on a single peril (drought) with a mechanism centred on individual owners and animals reflect how pastoralists themselves respond to drought? Based on research from 2019 to 2022 in

Borana, southern Ethiopia (Taye, 2022), this chapter investigates how pastoralists respond to drought-related uncertainties and asks whether insurance benefits impoverished pastoralists as intended.

Insurance and its assumptions

Insurance as a mode of social protection emerged as a result of the demise of cooperative forms of social support in Europe in the eighteenth and nineteenth centuries, as well as an increase in individualization and the market orientation of welfare. As a result of the expansion of the welfare state during the twentieth century, insurance became more socialized in state-based support mechanisms (Ewald, 2020). In recent years, insurance has been governed by the neoliberal doctrine of regulating at a distance through the market and is part of a larger trend of financialization, including of disaster risk (Johnson, 2020, 2021). Risk management through insurance, according to Dorfman (1998: 2), is 'the art and science of forecasting probable losses and formulating an efficient plan to survive them'. It is therefore argued that insurance is an important type of risk management, benefiting both agricultural production and the economy in general. However, as the literature on the social and political implications of insurance demonstrates, such technological, financialized, and market-based development interventions are never neutral or without social and political ramifications (Ewald, 1991; Isakson 2014; Johnson, 2020).

Unlike standard insurance, index-based insurance does not directly cover actual losses caused by a natural disaster such as drought. In traditional indemnity insurance, risks are identified, measured, anticipated, and calculated, and a loss is indemnified. As a result, the detected hazard and the damaged/lost asset/property have a direct relationship. Index-based insurance, on the other hand, employs an external indicator index (such as rainfall, temperature, or vegetation cover) to measure, predict, and indemnify a danger (such as the drought that causes crop or pasture loss). The advantages of index-based insurance over conventional indemnity-based insurance are threefold: 1) the transaction costs of verifying damages/losses are lowered; 2) the problem of 'moral hazard'[1] is minimized or resolved; and 3) the problem of 'adverse selection' is removed[2] (Isakson, 2015; Janzen et al., 2016).

For IBLI, satellite technology is used to monitor the availability of pasture vegetation. Vegetation cover in a given area is quantified and converted to an index using the Normalized Difference Vegetation Index (NDVI). The payout to insured pastoralists in a given area is determined by the value of the index throughout a season, with measurements made twice during the rainy season before the dry period begins. IBLI pays individual pastoralists prior to the occurrence of a drought, when the forage condition falls below an agreed threshold by comparing it to that in the previous 20 years.

All pastoralists in an insurance cluster pay the same premium per animal and receive the same rate of payment. The payments (if they occur) are intended to cover the cost of keeping an insured animal alive during the dry months.

As a result, the payouts are a monetized estimate of feed, water, and veterinary service expenses per insured animal during that particular season.

Several assumptions are built into the IBLI model:

- First, the model assumes that the distribution of rainfall is highly correlated with pasture availability and so drought risk. Further, it is assumed that pastoralists in a specific area have comparable pasture access, and their mobility is constrained to that area – the satellite vegetation monitoring is linked to a specific area only.
- Second, it is assumed that all pastoralists in an area are equally affected by the risk of forage scarcity and so drought. All forms of livestock are standardized into a single unit known as TLU (Tropical Livestock Unit) to monetize the effects of drought on livestock; thus, cattle and sheep (grazers) and camels and goats (browsers) are standardized and presumed to be equally affected by forage scarcity in a region.
- Third, IBLI is made commercially available to the most vulnerable pastoralists, and it is assumed that this will help them safeguard their assets (livestock) from the effects of drought, thereby minimizing the risk of sliding into poverty. The insurance model therefore assumes that animals are owned and held individually, that responses to drought are individualized, and that such responses are assisted through payouts that help protect an individual's household assets.

Overall, IBLI assumes a defined risk (a peril), which can be calculated and marketized with insurance sold as a product to individuals who can protect their assets if the threshold conditions in a specified area are met. Insurance, by definition, cannot address uncertainty, where future likelihoods of events cannot be calculated or predicted, and so products such as IBLI carry with them certain assumptions about how droughts occur and how pastoralists respond.

This chapter explores these assumptions with empirical information from two sites in Borana, southern Ethiopia, comparing the responses to drought risk and contrasting the socio-economic backgrounds of insured and uninsured households. The chapter shows how insurance, if purchased, is always combined with other responses and, in this way, pastoralists are able to respond to uncertainties, not just defined, calculable risks.

Borana, southern Ethiopia

Pastoralism in southern Ethiopia, as in other dryland areas, is an economic activity, a land-use system, a socio-cultural system, and a way of life in general (Coppock, 1994; Bassi, 2005). More than 14 per cent of Ethiopia's 110 million people reside in pastoral areas (CSA, 2013). Furthermore, pastoral production contributes significantly to the national economy by providing 40 per cent of cattle, 75 per cent of goats, 25 per cent of sheep, 20 per cent of equines, and 100 per cent of camels (FDRE/MoFED, 2017). Approximately 60 per cent of Ethiopia's geographical area is considered to be under pastoral production (Gebremeskel et al., 2019).

IBLI is currently being implemented in a number of pastoral areas in Ethiopia, including in Oromia Region (Borana Zone, 13 districts and Hararghe Zone, 1 district), Somali Region (10 districts), and Southern Region (1 district). IBLI is provided as part of the state–donor intervention and offers between 70 per cent and full subsidy in all areas except Borana where IBLI has been sold commercially since 2015. Oromia Insurance SC (OIC), the sole underwriter of the insurance product in Borana, had sold over 20,000 policies by 2021. The total value insured was 120 million Birr (around US$5 million) (OIC, 2019).

For this study, two insurance clusters were chosen from the 24 in Borana Zone. These were an extensive pastoral area (Dire) and an agro-pastoral area (Gomole) (Figure 7.1). The following criteria were used to identify these two clusters: agroecology (pastoral and agro-pastoral); a substantial number of insurance sales (insured households more than 5 per cent of total); and the absence of discounts or insurance subsidies. In both sites, livestock are the primary source of livelihood and revenue, although farming is becoming increasingly important in Gomole. A mixed methods approach was used in the study, combining quantitative (stratified household survey, N = 300) and qualitative (case studies (72), focus group discussions (12), ethnography (18), photovoice (18), and elite interviews (16)) methodologies. A total of 530 households and individuals were chosen, based on their insurance interaction (insured, dropouts, and uninsured), their wealth status (poor, medium-wealth, and rich), as well as gender, age, and location.

Comparing insured and uninsured pastoralists[3]

When IBLI was first introduced in Ethiopia, it was assumed that not everyone would participate, even on a commercial basis. Even though it was offered to everyone, it was assumed that some of the richer pastoralists would not take it up as they could self-insure without buying insurance. As a result, it was viewed as an intervention in support of vulnerable people.

Against expectations, the largest livestock owners (nearly all men) bought insurance in both locations (see Table 7.1). In Dire, insured families own twice as many animals (both large and small stock) as uninsured ones. Similarly, insured households in Gomole own 40 per cent more animals (both large and small stock) than uninsured households. Furthermore, insurance policyholders are more likely to own farmland as well as houses in town, indicating an overall higher socio-economic standing. The utilization of crop residue for animal feed is related to farmland size, which is larger among insured households and falls among dropouts and uninsured households.

Insurance was taken up by richer pastoralists, but how important was it for others? The data show how the majority of female-headed households are either uninsured or have dropped out of the insurance scheme. Dropouts are often under the age of 40, with a higher proportion being female.

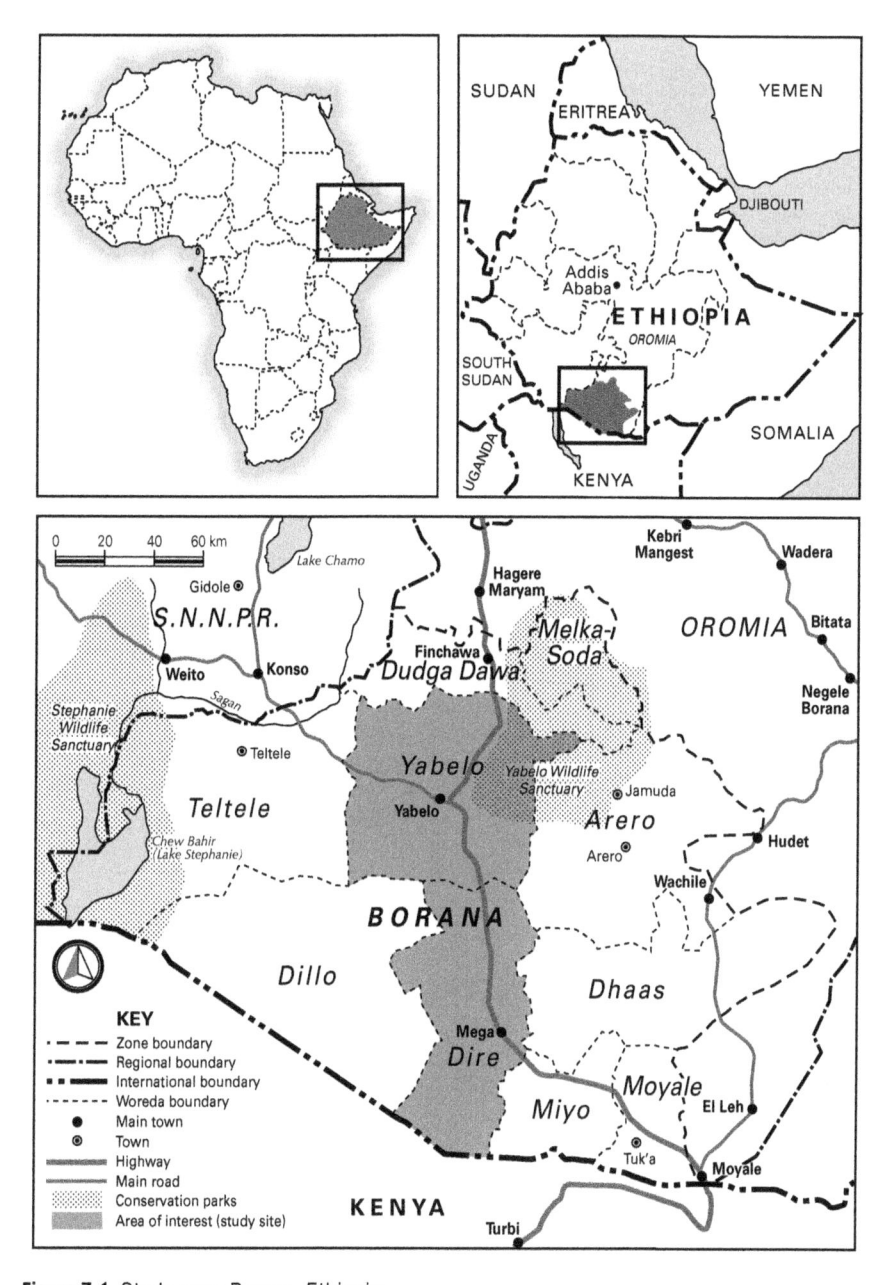

Figure 7.1 Study area: Borana, Ethiopia

Wealthy households invested 61 per cent more cash in buying insurance products than medium-wealth pastoralists and 244 per cent more than poor insured households. The total sum insured per household differs across sites, with the average being US$122 in Dire and only US$55 in Gomole.

Table 7.1 Key features of the population from Gomole and Dire (100 households in each category)

Selected indicators	Gomole			Dire		
	Active	Dropouts	Uninsured	Active	Dropouts	Uninsured
Female-headed households (from total respondents in %)	2.6	11.9	8.3	8.2	17.1	20.0
Gender (0 = male and 1 = female)	0.28	0.42	0.28	0.29	0.41	0.47
Age of household head (years)	44.0	39.1	41.3	46.6	37.7	43.6
Large stock (cattle and camels) owned (TLU)	19.9	13.2	14.2	30.3	16.7	14.6
Small stock (goats and sheep) owned (headcount)	42.1	18.0	26.0	42.8	37.5	27.9
Crop farm area (ha)	2.7	2.3	2.4	0.7	0.5	0.6
Houses owned in town (count)	0.5	0.3	0.3	0.4	0.2	0.2
% source of livelihood from livestock	50.6	47.4	38.2	76	73.3	67.5
% income from livestock	53.9	35.6	38.3	86.9	95.1	90
% income from crops	41.0	54.2	51.7	0.0	0.0	2.5
% pasture from *kallo* (community)	7.7	11.9	11.7	27.9	31.8	47.5
% pasture from crop residue	43.6	42.4	33.3	8.2	2.4	2.5

Pastoralists with many livestock have a limited (targeted) source of livelihood and income, yet multiple pathways to accumulate assets (farmland, houses in towns, etc.) and secure livestock inputs (pasture and water). The socio-demographic (gender and age) and economic (source of income and wealth) patterns suggest that wealthy adult males dominate insured households. Younger and female pastoralists, on the other hand, are dropouts with middle-class incomes.

Combining insurance with local responses

In 2019, a severe drought struck Borana, and the insurance underwriter paid out US$170,000 to 3,000 policyholders. In Dire and Gomole, 200 households received compensation, half of which were chosen for this study. Because the payment was made after the primary rainy season (March–June) failed, the mix of insured and uninsured responses could be determined. How did pastoralists – differentiated by location, wealth, and gender – combine insurance with other responses in order to offset the impacts of drought? Across the

Table 7.2 Combining insurance with local responses in 2019 (per cent)

Location	Insurance category	Wealth category	Increasing the purchase of feed	Consuming less food (daily) – either quality or quantity	Expanding private enclosures	Migrating to common areas	Starting farming
Dire	Insured	Poor	45.7	54.3	32.9	58.6	37.1
		Medium	46.7	53.3	34.5	57.3	50.0
		Rich	60.6	49.4	35.0	70.6	45.8
	Uninsured	Poor	42.2	65.6	32.2	43.3	32.2
		Medium	37.3	61.3	26.7	62.7	33.3
		Rich	40.0	51.4	28.6	74.3	40.0
Gomole	Insured	Poor	47.1	68.9	25.0	40.0	84.3
		Medium	59.6	56.8	32.6	38.5	90.8
		Rich	66.4	52.5	33.8	42.4	89.4
	Uninsured	Poor	40.7	81.7	22.2	34.2	73.3
		Medium	42.0	70.0	24.6	51.6	80.0
		Rich	50.6	63.7	23.5	48.2	70.6

sample households, 22 different responses were identified, 12 of them combined with index-based insurance. The following sections explore some of the main response strategies, contrasting those of insured and uninsured households (Table 7.2).

Purchase of animal feed

While insured families spent more than uninsured ones on average, there are significant variations between income levels and geographic regions. While low-income households in both locations spent more on animal feed overall, the spending gap between the insured and uninsured narrows when compared to those of middle- or upper-income status. The wealthiest families in Dire showed the largest difference between insured and uninsured households, with insured individuals spending 20 per cent more than uninsured families with the same income. In Gomole, insured households spent substantially more on feed than uninsured households.

Changing daily food consumption (quantity/quality)

Uninsured pastoralists practised this response strategy more than insured households in both sites; however, the application of this strategy is greatly determined by wealth status. Reducing food consumption is often used by low-income, uninsured families (65.6 per cent in Dire and 81.7 per cent in

Gomole). The rate decreases to 61.3 per cent in Dire and 70 per cent in Gomole for uninsured, medium-wealth groups. Uninsured wealthy pastoralists are the least likely to practise consumption-smoothing, although a significant number still do (51.4 per cent in Dire and 63.7 per cent in Gomole).

The money from payouts received by medium and poor households in 2019 was spent on food more than anything else. Those most likely to report this were the medium wealthy in Dire (27.7 per cent) and poor households in Gomole (28.9 per cent). Combining insurance payouts with food purchase, however, follows a different rationale. Poorer insured pastoralists bought grains for food once the payout had been made, with rice being the most common food item. A family of 10 can be fed with a kilogramme of rice, which costs $0.4. 'Rice is simple to prepare. You boil it, add salt, and serve it. You can save the milk you're selling by doing so. A litre of milk costs the same as a kilo of rice, yet serving 10 people requires nearly three litres', Negele – a young, insured poor pastoralist in Gomole – explained in September 2020. Although the payouts to the poor and some middle-class pastoralists were modest, they went a long way towards acquiring food because they insured a limited number of their livestock. In August 2020, Qaballe Dida from Gomole, who received US$15, said she was able to fulfil her food needs for over 17 days with the money she received.

Expanding private enclosures

The expansion of private land for grazing is an important response for some pastoralists. Particularly in Gomole, the form of ownership has shifted dramatically from communal to private holdings in recent years. Poor, uninsured households in Dire (32.2 per cent) and Gomole (22.2 per cent) are the least likely to expand private enclosures compared to the other two wealth groups. The wealthy in Gomole violate local rules by incorporating community land into their private plots, and insurance payouts incentivize pastoralists to expand private enclosures ahead of the rainy season.

Migration to common areas

Migration is an important strategy for pastoralists in Borana (Coppock, 2016). When pastoralists perceive pasture conditions to be deteriorating, they move their animals to common areas. More poor insured households (58.6 per cent) migrate to common areas that are closer to their basecamp than uninsured households (43.3 per cent). However, insured medium (57.3 per cent) and wealthy (70.6 per cent) pastoralists in Dire practise it less than uninsured households of the same wealth groups (62.7 per cent and 74.3 per cent respectively). The trends are similar in Gomole. However, when a comparison is made among the three wealth groups within each cluster, in both sites, migrating to nearby areas is more often practised by insured wealthy pastoralists (Gomole, 42.4 per cent and Dire, 70.6 per

Figure 7.2 A pastoralist fencing grazing land. When it starts raining, pastoralists enclose areas that are closer to their villages to improve grass growth for livestock feeding in the dry season.
Credit: Galmo (from the photovoice exercise).

cent) than the rest, as gaining access to land in other areas requires good connections and sufficient labour to move herds.

Although mobility is an inherent part of drought response (see Chapter 3), insurance payouts discourage pastoralists' movements. Rather than investing in collective responses – pooling livestock and labour for movement over longer distances – they instead seek to purchase water and feed, establish their own enclosures, or invest in farming through more individualized responses. In so doing, they undermine the community-level responses to drought centred on mobility, affecting other pastoralists' ability to respond to uncertainty.

Expanding farming

Despite the fact that farming is not common in Dire, with the average landholdings being less than a hectare, the expectation of low rainfall increases the likelihood of cropping among insured households compared to the uninsured. Medium-wealth groups combine farming with insurance more than the other two groups. In Gomole, wealthier households combine farming at a higher level than the other two: 89.4 per cent compared to the 70.6 per cent of uninsured households.

In sum, Borana pastoralists combine insurance payouts with a range of diverse responses. This changes according to wealth and location. Insurance payouts are used in a variety of ways. For some, they provide additional cash to purchase feed (or water, or veterinary services) and preserve their livestock assets, as intended by insurance promoters. Such responses, however, are concentrated among the wealthier pastoralists, particularly in the more pastoral area of Dire, where other options are limited.

Insurance payouts may also provoke unplanned-for actions, such as triggering investments in expanding private enclosures or the withdrawal of commitments to collective movement of herds. These may have negative knock-on consequences for others. For poorer households (and indeed for many others), the dominant drought response is to reduce food consumption, and insurance payouts may help offset this by providing households with cash to buy food. While this offsets the need for distress sales of livestock, the payouts are not focused on protecting livestock assets in advance of drought mortalities as intended. For insured pastoralists, insurance payouts are seen as part of a suite of responses, and often not the most important one. Shifting to non-pastoral livelihoods – including farming or trading – is often more important, especially in sites such as Gomole where other options are available.

For both richer and poorer pastoralists without insurance – indeed the majority of households in both Borana sites – it is this juggling of different options over time that is important. As a drought unfolds (and they are all different, affecting pasture, water sources, and livestock in different ways), pastoralists must combine responses so as to confront uncertainty. This cannot be defined in advance and requires skilled adaptation, drawing on long experience, collective knowledge, and networks of support. Insurance may be part of this portfolio of responses for some, but it is far from a silver-bullet solution to the challenges of drought response in the pastoral drylands of Borana.

Conclusion

According to its promoters, livestock insurance is supposed to strengthen the resilience of the most vulnerable pastoralists against drought, with the payouts intended to help keep core breeding livestock alive during climate-induced droughts. While insurance is strongly associated with the purchase of feed during drought periods, it is more important for richer, male pastoralists who are able to combine insurance with other responses. Wealthier pastoralists invest in productive responses that protect their herds and flocks, even enhancing their asset holdings during drought periods by selling older animals and purchasing younger ones.

Those few poorer and medium-wealth households who take out insurance are able to reduce distress sales, and do not reduce meals as much as those who do not have insurance, as the cash payouts can be deployed to purchase food for human use. Other poorer pastoralists may not be able to afford the commercial premiums for insurance and with fewer assets of their own must rely on other responses to drought, including more collective responses based on redistribution and sharing (see Chapter 6).

Most poorer pastoralists do not have access to individualized enclosed pastures to the extent that richer households do, so they must move animals during drought as part of a collective response. However, some medium-wealth households with insurance are buying water and feed during extreme stress periods, and many have also diversified into farming, especially in Gomole, reducing incentives to move and engage in collective responses, a shift in strategy which especially affects the poor.

Multiple factors therefore influence how pastoralists respond to drought, including economic status, gender, age, and location, as well as how insurance is combined with other responses. If insurance is purchased, it fits into a wider set of responses, which differ across households and locations. Insurance is not the 'pro-poor' solution to drought risk that some envisaged, although it can help in different ways if combined with other strategies, and become part of a wider portfolio of responses.

However, as the study showed, the assumptions underlying index-based insurance are not upheld in the Borana context. Insurance is directed to individual animals and households, but there are more collective responses that are also important, including movement. Insurance assumes that risks can be calculated in advance of drought onset based on indices of vegetation, but droughts affect different livestock in different ways, and people make use of rangelands in flexible ways that go beyond a delineated area.

Pastoralists' responses must take account of how droughts unfold over time and space, and how they affect animals in different ways. These impacts cannot be predicted, and so pastoralists must always adapt incrementally through a season. A single payout may help, but it must be combined with other responses – whether management of animals, or diversification of income sources, or changing food consumption patterns. In other words, it is uncertainty – where the likelihood of events is unknown – that is being addressed, not risk. As a market tool for addressing a singular peril, insurance is a risk management tool and cannot address uncertainty. Instead, local, contingent, adaptive responses, often rooted in collective responses, must come into play to complement or replace insurance if drought challenges are to be addressed (see Chapter 6).

Technocratic assumptions in insurance design are based on static derivatives, which are especially problematic in areas where land use, agricultural production, and socio-institutional systems are constantly changing, making forecasting difficult (see Chapter 1). Insurance instruments like IBLI must therefore be integrated within such social, cultural, and economic environments (cf. Ewald 1991, 2020), avoiding the dangers of such individualized market instruments displacing or contradicting more social forms of support.

Insurance must become embedded in wider social relations (such as gender dynamics), institutional arrangements (such as mobility and pastoral resource governance), economic livelihood strategies, and political dynamics in pastoral systems. As a market-based, individualized approach, insurance is not in any way superior to what are deemed 'traditional coping mechanisms', as is sometimes suggested. Indeed, quite the opposite: it is such embedded local responses that make it possible for insurance to function as a complement to collective, communal forms of response grounded in forms of local solidarity and moral economy.

Notes

1. Moral hazard can be understood as the incentive of an insured person to make unusual risk ventures.
2. Adverse selection is a situation when only one of the parties (seller or buyer of insurance) has knowledge or information about the risk to be insured. In other words, it is information asymmetry/failure.
3. Insured/active policyholder – a pastoralist/household with active insurance coverage during 2019/2020 and for one whole year before the time of the study. Dropout – a pastoralist who had purchased the nsurance product and then left the insurance scheme during the study period. Uninsured/non-policyholder – a pastoralist with no history of investment in livestock insurance.

References

Barnett, B.J., Barrett, C.B. and Skees, A.J. (2008) 'Poverty traps and index-based risk transfer products', *World Development* 36: 1766–85 <https://doi.org/10.1016/j.worlddev.2007.10.016>.

Bassi, M. (2005) *Decisions in the Shade* (C. Salvadori, Trans.). The Red Sea Press, Asmara.

Carter, M.R., Little, P.D., Mogues, T. and Negatu, W. (2007) 'Poverty traps and natural disasters in Ethiopia and Honduras', *World Development* 35: 835–56. <https://doi.org/10.1016/j.worlddev.2006.09.010>.

Catley, A., Lind, J. and Scoones, I. (2013) *Pastoralism and Development in Africa: Dynamic Change at the Margins*, Routledge, London.

Chantarat, S., Mude, A.G., Barrett, C.B. and Turvey, C.G. (2009) *The Performance of Index-Based Livestock Insurance: Ex Ante Assessment in the Presence of a Poverty Trap*. Available from: <https://papers.ssrn.com/sol3/papers.cfm?abstract_id=1844751> [accessed 22 November 2022].

Coppock, L. (1994) *The Borana Plateau of Southern Ethiopia: Synthesis of Pastoral Research Development, and Change. 1980–1991*. International Livestock Centre for Africa, Addis Ababa.

Coppock, L. (2016) Pastoral system dynamics and environmental change on Ethiopia's North-Central Borana Plateau—Influences of livestock development and policy, in R. Behnke, and M. Mortimore (eds), *The End of Desertification?* pp. 327–62. Springer-Verlag, Berlin Heidelberg.

CSA (2013) *Population Projections for Ethiopia, 2007–2037*. Central Statistical Agency, Addis Ababa.

Dorfman, M.S (1998) *Introduction to Risk Management and Insurance*, Prentice Hall, Saddle River NJ.

Ewald, F. (1991) 'Insurance and risk', in G. Burchell, C. Gordon and P. Miller (eds), *The Foucault Effect*, pp. 197–210, University of Chicago Press, Chicago IL.

Ewald, F. (2020) *The Birth of Solidarity: The History of the French Welfare State*, Duke University Press, Durham NC <https://doi.org/10.2307/j.ctv1168bj4>.

FDRE/MoFED (2017) *Country Annual Report 2015/16*. Ministry of Finance and Economic Development, Addis Ababa.

Gebremeskel, G. et al. (2019) 'Droughts in East Africa: causes, impacts and resilience' *Earth Science Reviews* 193: 146–61, <https://doi.org/10.1016/j.earscirev.2019.04.015>.

Homewood, K. (2008) *Ecology of African Pastoralist Societies*, James Currey, Oxford.

Isakson, S.R. (2014) 'Food and finance: the financial transformation of agro-food supply chains', *Journal of Peasant Studies* 41: 749–75.

Isakson, S.R. (2015) 'Small farmer vulnerability and climate risk: index insurance as a financial fix', *Canadian Food Studies* 2: 267–77 <https://doi.org/10.15353/cfs-rcea.v2i2.109>.

Janzen, S.A., Jensen, N.D. and Mude, A.G. (2016) 'Targeted social protection in a pastoralist economy: case study from Kenya', *Revue Scientifique et Technique-Office International Des Epizooties* 35: 587–96 <https://dx.doi.org/10.20506/rst.35.2.2543>.

Johnson, L. (2020) 'Sharing risks or proliferating uncertainties? Insurance, disaster and development', in I. Scoones and A. Stirling (eds), *The Politics of Uncertainty: Challenges of Transformation*, pp. 45–57, Routledge, London. <https://doi.org/10.4324/9781003023845>.

Johnson, L. (2021) 'Rescaling index insurance for climate and development in Africa', *Economy and Society* 50: 248–74 <https://doi.org/10.1080/03085147.2020.1853364>.

Lind, J., Okenwa, D. and Scoones, I. (2020) 'The politics of land, resources and investment in Eastern Africa's pastoral drylands', in J. Lind, D. Okenwa and I. Scoones (eds), *Land, Investment and Politics: Reconfiguring East Africa's Pastoral Drylands*, pp. 1–32, James Currey, Woodbridge <https://doi.org/10.2307/j.ctvxhrjct>.

Little, P., Mahmoud, H., Tiki, W. and Debsu, D. (2012) *Climate Variability, Pastoralism, and Commodity Chains in Climate Variability, Pastoralism, and Commodity Chains in Ethiopia and Kenya*. Adapting Livestock to Climate Change Collaborative Research Support Program, Colorado State University, Boulder.

McPeak, J., Little, P.D. and Doss, C.R. (2012) *Risk and Social Change in an African Rural Economy: Livelihoods in Pastoralist Communities*, Routledge, London.

OIC (Oromia Insurance Company SC) (2019) *High-Level Policy and Technical Workshop on Index-Based Livestock Insurance in the IGAD Region: A Report of the High-Level Policy and Technical Workshop on Index-Based Livestock Insurance in the IGAD Region*, ILRI, Addis Ababa.

Taye, M. (2022) 'Financialisation of Risk among the Borana Pastoralists of Ethiopia: Practices of Integrating Livestock Insurance in Responding to Risk', PhD dissertation, Institute of Development Studies, University of Sussex, Brighton <https://sro.sussex.ac.uk/id/eprint/109001/>.

CHAPTER 8
Confronting uncertainties in southern Tunisia: the role of migration and collective resource management

Linda Pappagallo

Introduction

Parts of the southern drylands of Tunisia in the region of Tataouine, bordering Libya and Algeria, are like inhabited lunar-scapes. These dry, highland areas have low and variable rainfall, yet hundreds of villages populate this challenging environment. The *jbeili* (or *djebalia, jbaliya*, mountain people) identify as Amazigh, or Arabized Berbers, and until the late 1990s largely lived in trogloditic abodes carved deep into the walls of the rocky canyons within the *ksours*, or fortified villages. Douiret is one of these villages. It is found on the crest of the Dahar ridge, where a plateau of 80 km rises gently towards the east at the edge of the Djeffara Plains (Figures 8.1 and 8.2).

Encountering this particularly harsh landscape may give the impression that these villages are isolated and marginal communities. However, the long migration history of the *jbeili* reveals how mobility through particularly strong family ties allow Douiris to remain at the centre of much wider and complex relationships that have been established across space and generations. Today, the link between human migration, livestock-keeping (mostly of sheep), and territorial management (of grazing) in managing uncertainties is an important one. The socio-economic options generated through this networked community allow the inhabitants of Douiret to live with unfolding uncertainties linked to variable precipitation, unemployment, inflation, and wider challenges of food insecurity. This chapter therefore discusses how human migration and collective resource management are an integral part

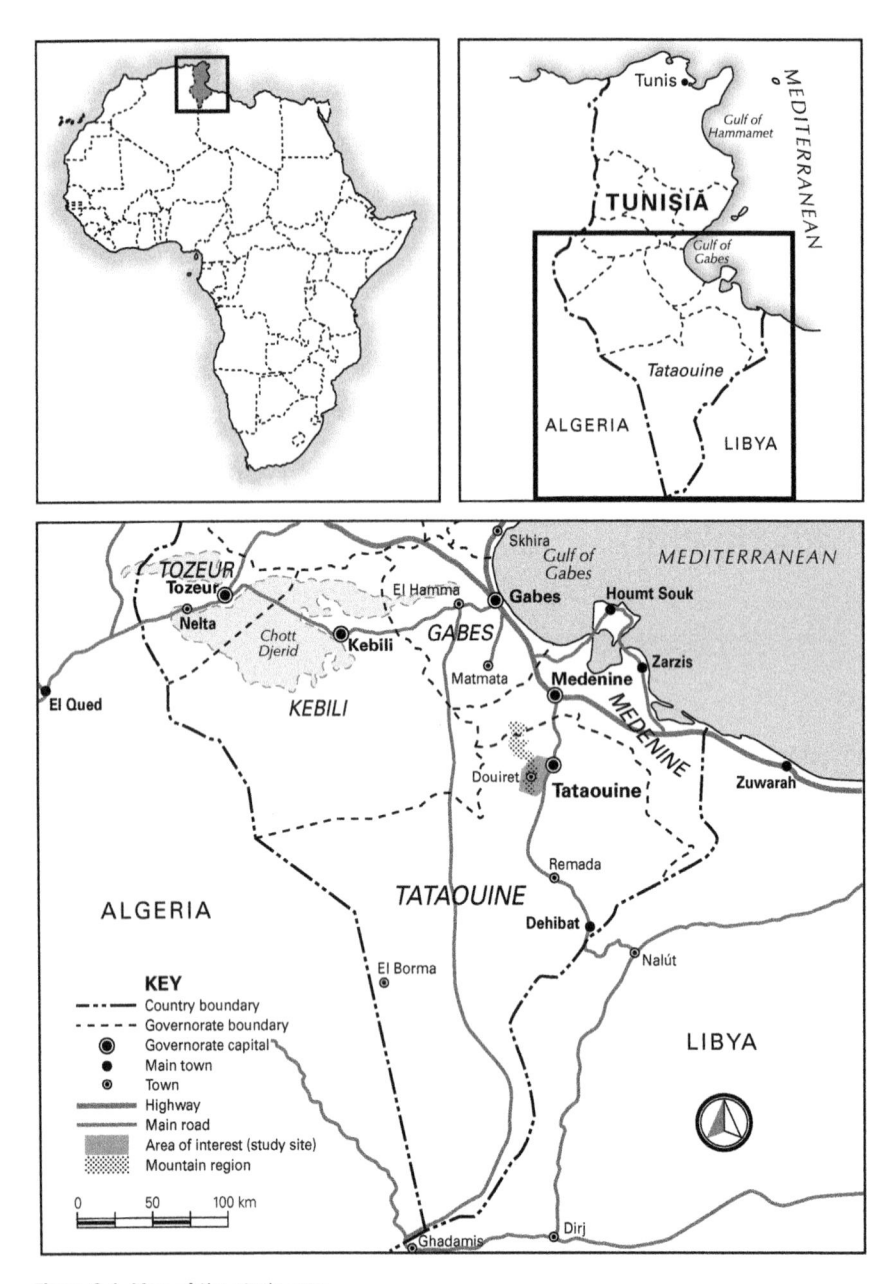

Figure 8.1 Map of the study area

of the adaptive strategies pastoralists use to meet changing conditions in this frontier economy.

Over centuries, the *jbeilis* have managed uncertainties with great skill. Geographically, Douiret is surrounded by *wadis* (dry valleys that become riverbeds during the rainy season), marked by south-west facing *jessours*, a

Figure 8.2 The old town of Douiret, with its troglodytic abodes, became largely unpopulated by the late 1980s as villagers moved to 'new' Douiret or elsewhere.
Credit: Linda Pappagallo.

series of terraces formed behind *tabias* (water and soil catchment barrages built from earth and stone). (Figure 8.3).

These ingenious structures domesticate a difficult terrain to increase livelihood options. The *jessours* increase soil moisture for crops, ensure groundwater recharge through the infiltration in the terraces, and protect downstream infrastructure by flood control. For Douiris, better pastures, olive tree plantations, and partially cultivated cereals in the *jessours* remain key for their livelihoods.

For livestock owners, the *jessours* offer high-quality grazing in the mosaic of grazing options. By relying on a spectrum of relations and land tenure regimes, from open pastures and commons belonging to and managed by clans (called *arch*) to more private (but unfenced and accessible) land like that in the *jessours*, livestock owners are continuously finding ways to manage their flocks' nutritional requirements by tracking patchy rainfall distribution. This chapter asks how livestock owners in Douiret overcome the challenges posed by multiple uncertainties linked to shifting environmental, economic, and socio-political conditions in a highly challenging terrain.

Two key aspects emerge. The first is the long and particular migration history that marks Douiret's socio-economic establishment outside of the village, and the second is the evolution of collective herding arrangements such as the *khlata* described below. Migration and the resulting 'absences' of various family members, combined with the pooling of resources, are therefore two strategies that pastoralists in Douiret use to manage uncertainties (Pappagallo, 2022).

Figure 8.3 The *jessours* help with water and soil management in order to increase livelihood options.
Credit: Linda Pappagallo

Migration and 'absence'

Today, Douiret has a total population of around 90,000, but only between 700 and 1,000 live more permanently in Douiret. The majority (around 20,000) are in Tunis, the capital of Tunisia, some 500 km away, specifically in the Medina (the old part of the city) where 'little islands' of Douiret have developed over time in specific neighbourhoods such as Hafsia, Baba Jezira, and Bab Jedid through processes of aggregation and spatially stretched social relations (Prost, 1955). As Ali, a Douiri who grew up in Tunis noted in a February 2020 interview, 'We are not in the village, we re-create the village so we are not far from the village' and this is a specific trait that marks the historic relationship between Douiret in the south and Douiret in the north, in Tunis.

The migration patterns of the *jbeili* to the north are longstanding, with particular rhythms. Each village has typically (though increasingly less so) operated in a well-defined and established economic space. In the case of Douiret, for example, this has developed especially through Douiris working as porters in the central market, bakers, and newsagents. In the 1960s, migration corridors evolved to France, Libya, and recently to Canada. The enduring significance of these networks is particularly crucial in such harsh environments where networks and the extended family continue to form the social architecture through which Douiris outside Douiret remain connected and support livelihoods and incomes inside.

Migration is therefore one of the ways by which Douiris have compensated for the economic and environmental uncertainties of their territory; by creating a differentiated and 'pluriactive' socio-economic landscape where non-farm work, informal markets, migration, and remittances remain central to rural transformations, and where the household is an economic institution straddling multiple localities (Fréguin-Gresh et al., 2015).

Since a significant part of the population are scattered in different places and therefore 'absent' from Douiret, the agrarian political economy must be assessed by who is present and absent over time. This relationship is key to understanding how people deal with uncertainties in drylands and highlighting what patterns of accumulation of livestock (and other resources) are possible – and for whom. For livestock owners, for example, absence influences how social networks around livestock management operate, how identities and subjectivities around livestock husbandry are formed and maintained, and how remittances from outside the area provide the basis for investment in local flocks. In fact, in Douiret, having a flock is an extended family project: various members contribute to the survival and growth of the flock at different moments. As Hedi explained in March 2021:

> I was herding on my own. I cannot afford to pay a herder 1,000 dinar [US$325] while I sleep at home. I had to herd on my own for the first and second year and make some sacrifices. My brother helped me financially to be able to afford the herding costs later. I have two siblings who work in Tunis. In the beginning, you need to have someone who can help and support you financially, otherwise you cannot make it on your own due to the increasing costs and forage prices.

For young livestock owners like Hedi, the absence of various family members therefore also acts as collateral against inflation in the local economy along with other uncertainties. External flows of cash – such as remittances – help to support the flock and ensure that expensive feed can be bought during long periods of drought so that part of the flock does not need to be sold. Remittances also help to regulate the fast-paced changes in livestock and meat prices. In Douiret, livestock owners are operating in a border economy, a context where livestock markets are highly dynamic due to frontier trading with Libya. Douiris benefit from arbitrage in these interstitial spaces by acquiring animals cheaply from Libya, especially when there is conflict, and selling to Libyans when things are better there. Access to informal credit through family networks also allows livestock owners to buy cheaply and sell for a profit, and so address market uncertainties. In this sense, the absence of family members cannot be seen as a void but must be understood as including various forms of connection to the territory.

The gendered and age-specific aspects of absence are also important. Absences define trust relations across networks; they change divisions of labour and so redefine notions of femininity and masculinity and the organization of split households. One reading, for example, is that for every man that leaves, there are (several) women who stay in order to support and enable different departures. Sisters, wives, mothers, and other female relatives stay in Douiret to support social reproduction, to allow for dislocated accumulation possibilities, and also to continue to sustain agricultural production – either directly or through the management of people.

Zoglem, an elder who worked a lifetime in the central market as a porter in Tunis, affirmed this:

> If it wasn't for women, we wouldn't be able to do anything; women are everything. When I was in Tunis, I had a lot of livestock here and I relied on my wife. She ran the business, she bought hay for the sheep and managed the flock, and spent from her own money to keep things going. If I were to do it alone, I couldn't have achieved that. My daughters helped her as well; without the help of my daughters and my wife, I couldn't have raised livestock.

This is the story of many individuals who, leaving Douiret for work, have depended on elders, women, and children who remain behind for agricultural work, such as herding and olive-picking. However, times are changing. Zoglem is of a particular generation in which women were more involved in pastoral labour. But today, aspirations related to ideas of modernity and concepts of a 'modern woman' mean there is a certain stigma attached to women taking care of the flocks. As the roles and responsibilities of elders, women, and children shift, and as external family support is not accessible to all, livestock owners in Douiret have to find other ways to manage livestock that are less reliant on family labour and networks. In these cases, the role of social institutions in mediating alternative arrangements of herding labour become key.

Pooling resources and collective herding arrangements

Young, cash-strapped livestock owners, without migrant networks sending remittances, build their flocks by tapping into collective livestock management arrangements such as the *khlata*. The *khlata* literally means 'mixing' in Arabic. It describes a pooling process whereby different livestock owners pool their flocks (or mix them together) and at the same time share herding labour costs, as well as pooling other endowments, such as pastureland and water. Other aspects of the collective management of the flock – such as transportation, fodder, and water costs – are negotiated along with access to tractors, boreholes, or markets. The *khlata* is managed by a herder who, apart from herding, watering, and tending the animals, may also be accountable for tracking costs and revenues, negotiating access to resources such as veterinary services, feed, land, and water, and consulting with owners around decisions on the sale of livestock by following fertility rates, calving seasons, and fluctuating market prices for meat.

The benefits of entrusting a private flock to a *khlata* arrangement are therefore particularly evident for individuals who do not have access to family labour, credit, remittances, savings, or capital to pay for herding labour costs. Through the *khlata*, labour costs are negotiated and shared amongst associates. Usually, the *khlata* is composed of 8 to 15 members, and the *khlata* herd can reach up to 700 head. Through the *khlata*, the collective flock has the right to access to pasture that 'belongs' to each of the members' clans. This means that if there are 15 members, each with different endowments or access rights (including some with no access to land), the collective herd has access to a variety of grazing options through the *khlata*.

By pooling different and dispersed land endowments, livestock owners can use the *khlata* to overcome the requirement to individually negotiate resources and relationships that are important for the flock, and so address uncertainties. As Ben Brahim explained when interviewed in March 2021, 'For instance, if I want to go herding in Ouara, they would kick my flock out, they wouldn't allow me to grazethere [...] If I want to get access to it, I will have to make a *khlata* arrangement with someone there.' The Ouara are pastures in the plains, so to gain access Ben Brahim would have to build relationships and reciprocal arrangements through the *khlata*.

The *khlata* thus legitimizes access to land, and such pooling arrangements can also expand rights and subvert certain social expectations and rules about getting access to land. For example, the *khlata* alleviates the emotional pressure for younger siblings (constrained by gerontocratic power relations) or landless herders (constrained by the legacy of patron–client relations) to negotiate access to land, given the current place they hold within society. There are still tacit social pressures, norms, and expectations (such as trustworthiness) that determine whether you are able to enter a *khlata* arrangement and with whom, but these relationships are simplified by productive objectives and reciprocal interests. In this sense, the *khlata* does not simply guarantee access to land and resources, but it subverts customary rules that limit the capacity for individuals to obtain a degree of autonomy. This is another way in which collective pooling mechanisms and reciprocal arrangements not only address uncertainties in a resource-constrained setting but also act to redistribute unequal distributions of pasture or other endowments.

The heterogeneous qualities and patchy distribution of pasture in the Dahar, with changes within and between seasons, means that negotiating access to pasture and water is highly complex. For example, when certain rangelands are depleted, especially during dry spells or droughts, access to other rangelands must be negotiated within the farming community, a point highlighted by Ben Brahim:

> Everyone is struggling in their own way in Tataouine, and everyone knows each other. It's a small state. For example, as we know each other, you can come to ask me if you could herd in a specific part of my land

> since drought hit yours and not mine. I would then tell you to go ahead and I hope the best for them [...] It's only the farmer who would feel the struggle of another farmer [...] The farmers here know each other and deal with each other, that's it.

Although access to pasture depends on the bundle of social relations and the power of negotiation that the individual has within the community, when management is collectivized, the expertise and connections required to track pasture and rainfall are considerably simplified. This is because access to land becomes a collective problem with collective solutions. For those who need it, pooling access rights through the *khlata* can be a strategy to gain access to more and better land, therefore stretching the rainy season (Krätli, 2016) and helping to synchronize access to pasture with patchy rainfall distribution.

Since the early to mid-2000s, there has been a revival of the *khlata*, in part fuelled by returning migrants' decisions to (re)constitute flocks after having accumulated sufficient capital elsewhere, and in part as 'absent' workers in the city aspire to own livestock in their rural homes. These individuals are often labelled 'absentee livestock owners', although they remain very much connected and so are 'present' in the territory socially and economically. For example, by negotiating their presence in the pastoral system through the *khlata*, these individuals use the flock to claim a sort of territorial visibility. By entrusting their flocks to the *khlata*, and by paying a fixed price per head, they can ensure that their flock is being looked after and can access pasture through the collective flock – even though they may have lost their access rights in their absence and with time. The *khlata* model is convenient for absentees when flocks are relatively small – for example, less than 60 head – since hiring an individual waged herder would be too expensive, and it would require livestock owners to be more physically present.

In order to achieve faster rates of accumulation, optimal nutrition is important to support higher lambing rates, even during drought. Initially, maintaining low costs and building territorial knowledge and experience through the *khlata* can be beneficial, although it is widely recognized that collective arrangements may not result in the best nutrition for the animals. However, as flocks grow and the operation becomes more commercialized, some livestock owners may prefer to hire individual herders, as they can focus on the flock and manage its nutrition in a more highly attuned way. As the flock continues to grow, a young livestock owner wanting to establish himself will therefore often decide to leave the *khlata* arrangement. Size does matter and larger flocks are not suitably managed through a *khlata*. Interviewed in April 2021, a young livestock owner, Mousoud, stressed this:

> It is not profitable to engage in *khlata* when you have a large number of livestock. For instance, if I have 300 head and I mix it in the *khlata*, I would be spending the same amount as if I were to hire a

herder, but in the *khlata*, my livestock isn't eating that well since it's among a large number of other sheep. With a *khlata* of 700 and 800 head, my flock will not be eating that much. They can't even enter some private lands, while when my flock is on its own, it's always full [in terms of feed ...]I realized that I harmed my own livestock by mixing [...] even the flock will not be at ease as they would be following each other. In herding, they always say *kallel w dallel* [that the less the better]. *Khlata* is difficult, sometimes you can mix with a non-trustworthy person.[1]

To summarize, the harsh context of Douiret – with multiple, intersecting environmental and market uncertainties – dictates how resilience is built through migration, and the relationship between presence and absence. This allows for the taking of opportunities for accumulation elsewhere while remaining connected to one's territory of origin through collective pooling mechanisms, such as the *khlata*. Combining migration with collective pooling explains how pastoralists in Douiret navigate the uncertainties associated with such variable socio-ecological landscapes.

As the types of uncertainties shift with changing environmental and political-economic conditions, so the strategies and forms of institutions shift to respond to the new conditions. Understanding institutional adaptation and the evolution of the *khlata* thus further highlights the importance of adaptable and informal collective resource management.

The evolution of informal collective resource management institutions in southern Tunisia

The *khlata* is an institution that is responsive to the dry, uncertain conditions of the south of Tunisia, where rainfall is variable, services are limited, and the costs of purchasing fodder and water are high. Resource scarcities are therefore compensated for by reciprocal and collective arrangements.

It is the relative abundance or scarcity of resources that dictate livestock production and determine how collective pooling mechanisms are used. Changes in the politics of access to rangelands in the Dahar, and land fragmentation and privatization in the plains, for example, continuously reconfigure patterns and pressures of access to pastureland in the mountains, and consequently the configuration and use of the *khlata*.

While associates of the *khlata* in the past were typically from the same family (*arch*) or based around kin ties within Douiret, *khlatas* today are composed of a more diverse group of individuals, increasingly including members that are from outside Douiret but who have ties to the village. The trend of increasing heterogeneity in membership composition is a sign of institutional adaptability. The *khlata* increasingly includes relatives, neighbours, or strangers, people with different identities (non-Douiri), or non-kin (non-family members). It also includes a broader representation of classes, different forms of absence, different rationales and strategies, and different asset endowments. This adaptive mechanism relies on the flexible bundling and unbundling of social relations as Jalloul explains:

> The composition of the *khlata* has changed. Before the 1950s, flocks from Douiret could not mix with the flocks from other villages. Each village had their flocks and their rangelands. As the rangelands became less productive and patchy, the membership to the *khlata* began to evolve by including livestock owners fromother communities. This was to enable more extensive access to rangelands, which included 'poorer' as well as more fertile rangelands to feed the flocks better. This has happened since the 1980s (interview, April 2021).

Jalloul's reflection suggests that, as resources become more scarce or competitive, and as ties become more tenuous, extending beyond the narrow circle of the family or *arch*, the *khlata* arrangement, as a more collective form of associateship, becomes more important for accumulation. This seems to contradict two views: the view that suggests that increasing resource scarcity results in trends towards individualization and privatization (themselves seen as more efficient modes of production); and the view that collective action is most efficient only with small, homogeneous groups (Baland and Platteau, 1995). Instead, what pastoral production in Douiret shows is that migration coupled with collective forms of associateship, ones that are adapting to become increasingly heterogeneous, provides solutions to various forms of uncertainty.

Conclusion

Understanding how livestock-keepers in Douiret confront uncertainties requires using a larger spatial scale and looking at how forms of 'presence' are negotiated in the village territory and how 'absence' through migration generates connections and options. These connections provide links that support patterns of livestock accumulation in Douiret by, for example, sending remittances. But other less visible forms of connections are revealed by looking at, for example, how relationships vary with different age- and gender-specific roles, how rhythms of mobility influence the types of connections, or how types of networks that link the village and the site of migration influence patterns of mobility and engagement with territory. All these aspects clarify how absence entails various forms of connection that influence pastoral production. Practically, the next step is understanding how absence is mediated by individuals and families to ensure that livestock-keeping is maintained. What pastoralists in southern Tunisia teach us is that the *khlata* is an example of a collective institutional mechanism that helps to manage variability by sharing risks and options.

The adaptability of the *khlata* comes from its organizational framework: it is informal, negotiated, and depends on the mix of pooled endowments brought in by the different associates. It is in this sense that the *khlata* takes its strength from its collective structure. By having a heterogenous membership, with varied endowments (access to land, trucks, water, administrative connections), more options are created for the management of an individual's flock to deal with various climatic and socio-political uncertainties. The *khlata* is therefore an example of a collective institution and adaptive mechanism that provides the basis for resilient livelihoods in the drylands, at least for smaller flocks and especially for young, cash-strapped entrepreneurs, the landless, retirees, and absentees.

References

Baland, J.M. and Platteau, J.P. (1995) 'Does heterogeneity hinder collective action?' *Cahiers de La Faculte Economique et Sociales* 146.

Fréguin-Gresh, S., Cortes, G., Trousselle, A., Sourisseau, J.-M. and Guétat-Bernard, H. (2015). 'Le système familial multilocalisé. Proposition analytique et méthodologique pour interroger les liens entre migrations et développement rural au Sud', *Mondes En Développement* 4: 13–32 <https://doi.org/10.3917/med.172.0013>.

Krätli, S. (2016) 'Discontinuity in pastoral development: time to update the method', *Revue Scientifique et Technique de l'OIE* 35: 485–97 <https://doi.org/10.20506/rst.35.2.2528>.

Pappagallo, L. (2022) *"Partir Pour Rester?": To Leave in Order to Stay? The Role of Absence and Institutions in Accumulation by Pastoralists in Southern Tunisia*, doctoral thesis, Sussex University <https://sro.sussex.ac.uk/id/eprint/109122/>.

Prost, G. (1955) 'L'émigration chez les Matmata et les Ouderna (sud Tunisien)', *Les Cahiers de Tunisie: Revue de Sciences Humaines* 3: 316–25.

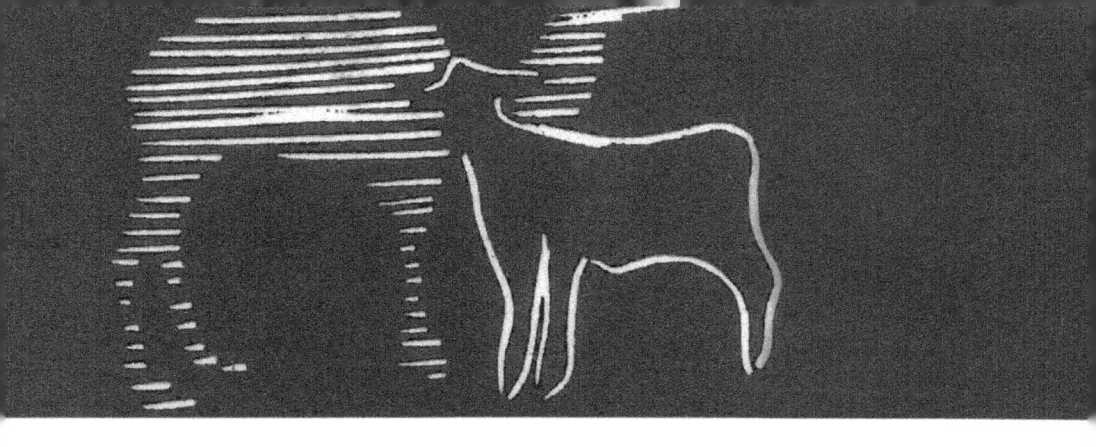

CHAPTER 9
Living with and from uncertainty: lessons from pastoralists for development

Ian Scoones and Michele Nori

Introduction

As we have seen across the cases explored in this book, the framing of much development policy in pastoral areas is centred on control, stability, and directed management towards a vision of modernization, usually associated with settlement and an agrarian or urban lifestyle. This runs counter to how pastoralists must live and produce in highly variable environments: by managing uncertainty, avoiding ignorance, and generating reliability. Contexts are fast-changing, and so are sources of variability and drivers of uncertainty, but pastoral practices remain rooted in the core principles of pastoralism discussed through the chapters (see also Nori, 2019).

This chapter offers a brief survey of pastoral policies across the regions where the case studies come from, drawing out the themes that are both common and contrasting. Unfortunately, the majority of existing policies run counter to the principles of pastoralism discussed throughout this book (see Chapter 1), acting to undermine pastoral practices rather than support them. Of course, development policies and interventions are not uniform, and there are many projects scattered across the world that do offer a perspective drawing on principles of openness, flexibility, and adaptation to generate reliable, robust, and resilient livelihoods in the pastoral rangelands. But these remain a minority. A World Bank report from 1993 reported that the pastoral sector had experienced the greatest concentration of failed development projects in the world (de Haan, 1993), and sadly the situation has not improved much since then. As we discuss below, what is essential, if an alternative development vision is to be realized, is to reverse the framing of standard, mainstream policy and so begin to 'see like a pastoralist' (Catley et al., 2012).

The chapters have examined various principles of pastoralism in different contexts, looking in turn at mobility, resource use, markets, insurance, moral economies, and the collective pooling of resources as a way of assuring accumulation even if herders are physically absent. These insights suggest a very different approach to framing policy for development. This chapter aims to pull these threads together, offering a future agenda for pastoral development that genuinely takes uncertainty seriously. And, as we argue in conclusion, this may offer some clues as to how everyone confronting a turbulent world can navigate uncertainty more successfully, requiring us all to abandon some of our long-held precepts about stability, control, and a particular form of linear, modernizing development (Scoones, 2019, 2022a; Scoones and Stirling, 2020).

Pastoral policies: a regional overview

What are some of the recurring themes of contemporary pastoral policy across a selection of world regions?[1] While there are obviously contextually specific differences relating to particular histories and political economies, there is a remarkable convergence across regions (see Nori, 2022 a–d for details of specific national and regional policies).

Pastoral policies in **China and the former Soviet republics of Central Asia** have been deeply affected by the shifts from state control and collectivization to an engagement with the market economy and individualization of tenure in the rangelands. Former state farms have been redistributed and cooperatives abandoned, and there is a commitment to investing in household-based livestock production, around which a set of modernization policies are promoted. In central Asian countries, public investments have faded in the transition to a market economy, giving rise to new patterns of territorial and social polarization (Kerven et al., 2021; Nori, 2022a).

China's high-profile Belt and Road Initiative has focused on developing transport and energy infrastructure, which in turn helps promote other investments including mining and crop-farming. While the shifts between regimes have been dramatic, in practice, pastoralists have always had to negotiate a way through centralized policies. In the state-controlled, collective era, there were elements of private and communal management and today, collective forms persist in new forms of hybrid rangeland governance that are compromises between centralized policies and local contexts (Tsering, 2022; Gongbuzeren et al., 2018; see also Chapter 4). Despite major changes in governance, the necessity of managing highly variable resources persists across regimes.

In **South Asia**, similar patterns of investment exist with, for example, huge industrial investments occurring in Kachchh in Gujarat, India following the 2001 earthquake. This comes on the back of longer-term development of irrigation infrastructure, including the contentious Sardar Sarovar Dam and canal system (see Chapter 3). As elsewhere in the region, state policy focuses almost exclusively on the intensification of agriculture, with pastoralism generally neglected. This echoes the Green Revolution period in India from the 1960s, when promotion of high-yielding crops were invested in. Negotiating access to increasingly intensified farms and brokering arrangements with crop producers, while moving across landscapes that are being transformed through industrial development and transport infrastructure, presents many challenges for pastoralists as they continue to respond to high environmental variability but now with new uncertainties combined (Maru, 2022, see Chapter 3).

In **Europe**, a different dynamic is evident. Peripheral mountainous and hilly areas inhabited by pastoralists are of interest to policymakers mostly for their environmental values, including biodiversity protection, but also as sites for solar and wind energy generation. On paper at least, pastoralists are appreciated as environmental managers of natural habitats and offered subsidies accordingly. In practice, however, the European Union's CAP favours consolidated farms with high input production systems and tight integration into markets. As a result, pastoral people have migrated, and local skilled herding labour has become short and been replaced by an immigrant workforce. Small farms close and many grazing lands are being abandoned as a result. Talk of subsidy reform has dominated policy debates across Europe over decades, but political allegiances, market interests, and commitments to certain classes of rural producers means that policy change has been slow (Simula, 2022; Nori, 2022b; see also Chapter 5).

In the **Middle East and North Africa**, subsidy regimes again affect pastoral systems. Policy efforts have been aimed at intensifying and stabilizing livestock production so as to serve the needs of a growing population. Subsidies for fodder importation, water development, community settlement, and market engagement combine to change pastoral production. This results in the uncontrolled growth of herd and flock sizes, the intensification of production, and patterns of over- and under-grazed areas, often resulting in serious land degradation (Nori, 2022c). High population growth and the development of urban economies in the region, accelerated through the oil boom, have triggered migratory flows to towns and cities and also to other countries, including to Europe and the Arabian Peninsula. Remittance money and hired herders are now integral parts of local livestock systems. This link between migration dynamics and pastoral production is important (Pappagallo, 2022; see also Chapter 8). As people migrate and become absentees (in many different forms), livestock and labour take on new roles, with significant implications for socio-political dynamics, including in class, gender, and generational terms.

Finally, in **sub-Saharan Africa**, many of the same patterns occur – constraints on mobility, rural depopulation, infrastructure investment, livestock commercialization, the increase in absentee owners/hired herders, and the diversification of livelihoods.[2] However, policy regimes differ across the continent. For instance, in Eastern Africa the African Union and the Intergovernmental Authority on Development have confirmed the importance of pastoralism for regional economies and the need for pastoral development.[3] Meanwhile, transhumance protocols have affirmed the importance of mobility, even if such frameworks focus on the control of movement and allow limited flexibility.[4] Regional and pan-African policy positions on pastoralism are increasingly positive, but this is not always reflected in national policies, which are much more mixed. National governments often see pastoralism as challenge for state-building, especially when pastoral areas are in borderlands where conflict is rife. However, where decentralization has taken place, local governments are more likely to be staffed by those from pastoral backgrounds and a more sympathetic position towards pastoral development may be adopted.

Many African countries are highly reliant on external investments and donor aid, particularly in the dryland pastoral areas where national funding is often limited. Policies encouraging external commercial investments have been encouraged as state resources decline. Investments in pastoral areas have included mining, alternative energy, conservation, irrigated farming, tourism, and more. Policies governing such investments have often been lacking, resulting in land, water, and green grabs in pastoral areas, and growing grievances among pastoral populations, along with disputes over land (Lind et al., 2020a; Nori, 2022d; Robinson and Flintan, 2022). Aid policies increasingly focus on humanitarian assistance, often resulting in a dependency on aid flows in some pastoral regions where drought strikes frequently, despite the rhetoric around resilience building, livelihood security, and social protection (Mohamed, 2022; Taye, 2022; see also Chapters 6 and 7).

While conflict has long been a feature of East and West African drylands, the intensity has increased, especially as conflicts have become bound up with global political-economic interests, and the availability of small arms has also increased following the wars in Libya and Somalia, for example (UNECA, 2017). This has resulted in a range of policy interventions to bring 'peace' and prevent the expansion of 'terrorism'. These militarized, securitized interventions, supported by international forces and heavily influenced by global geopolitics, are reshaping sub-Saharan African drylands. Pastoralists find it increasingly difficult to navigate such an uncertain and sometimes violent terrain, and may be forced to take sides (Benjaminsen and Ba, 2019). With pastoral areas deemed to be contributing to the 'fragility' of states and wider regions, restrictive interventions that undermine key pastoral strategies, notably movement and cross-border exchanges, are frequently implemented.

Narratives and policy change

Across these regions, there is therefore a remarkable convergence in policy thinking and practice, despite local particularities. We see a huge diversity of political systems – communism and state-direct capitalism, autocratic kingdoms, liberal democracies, and many variations in between – yet there are striking similarities in the way pastoral and dryland areas are viewed and how a future is envisioned within policymaking. In almost all cases, pastoralists are seen as threats to state stability, destroyers of the environment, and followers of 'backward' forms of production. They are therefore deemed to be in need of modernization and development of a particular type (Gabbert et al., 2021; Garcia et al., 2022).

How can a more positive view of pastoralism emerge? A first task is to address the misleading narratives that construct the current debate. Despite much accumulated evidence, such as that presented in this book, they are difficult to dislodge. They persist in institutionalized forms in government procedures, and the biases are frequently reiterated even by new generations of government officials and policymakers. This makes formal institutions and development agencies ill-equipped to deal with the complexity of pastoral systems. Bringing the themes discussed in the preceding chapters together, in the following sub-sections we briefly outline seven dominant narratives that frame contemporary policy in pastoral areas and highlight an alternative perspective that is more rooted in the uncertain, complex realities of pastoralism.

Environmental degradation

Many policies focus on the assumed patterns of environmental degradation in rangeland areas. A heavily grazed rangeland at the end of the dry season or during a drought may look in terrible shape to the untrained eye, but as soon as the rains come, the grassland bounces back. While this is not always the case – such as in the 'sacrifice zones' around watering points or near markets and towns – the ability of grasslands to recover is remarkable (Briske et al., 2003). However, too often, degradation and desertification assessments are derived from snapshot views. True degradation happens in rangelands where areas that act as key resources – such as wetlands or riverbanks – are removed from use, and the flexibility of movement and use of diverse areas is reduced. However, the frequent reaction to seeing rangelands is to proclaim them 'wastelands' in need of restoration. The assumption is that grasslands are a sub-climax vegetation type and if managed 'correctly', they would become covered in trees (Bond et al., 2019). However, tree planting in rangelands rarely makes sense. These are open ecosystems, maintained over millennia through a combination of grazing and fire (Bond 2019). Where environments are especially variable, they show non-equilibrium dynamics, never reaching a stable pattern, and with animal populations always below any fixed carrying capacity (Behnke et al., 1993; Swift, 1996).

'Degradation' in such a setting is less easy to define and requires an accurate assessment of baselines, time frames, and dynamic patterns (Scoones, 2022b). If livestock production on a rangeland declines over time, degradation may be occurring and restoration will be needed; but if grass species change and bare patches appear but livestock are still producing over the long term due to a range of practices responding to variability, the system is clearly resilient. By maintaining such open ecosystems, with transhumant routes connecting biodiverse areas, grazing preventing fire events by controlling biomass, and animals dispersing seeds to encourage plant diversity, pastoralism can contribute to conservation, enhancing ecosystem services (Yılmaz et al., 2019).[5] Taking account of these dimensions is therefore important in seeing pastoralism as a positive generator of environmental integrity and biodiversity, rather than as a source of degradation and desertification.

Climate change

Ruminant livestock – including cattle, camels, and smallstock – produce methane as part of the digestive process. Methane is a climate forcing gas that causes global warming, even if (unlike carbon dioxide) it breaks down in the atmosphere over around a decade. A strong narrative exists suggesting that all livestock production, including extensive livestock systems, needs to be reduced if the imperatives of the challenges of climate change are to be addressed (Steinfeld et al., 2006). This, it is argued, requires a major shift in global diets away from meat and milk towards plant-based products, or even lab-manufactured alternative proteins and the products of precision fermentation of bacteria.

However, different livestock systems have very diverse impacts and lumping them all together makes no sense (Houzer and Scoones, 2021). As governments, aid agencies, and environmental campaigners adopt a generalized anti-livestock message, often in the desperate scramble to meet targets and show climate-change commitments, the unintended impacts can be huge. Pastoral systems produce very few greenhouse gases compared to intensive, contained livestock systems; indeed, by some calculations, they are actually carbon neutral, as manure and urine is dispersed and incorporated in soils, sequestering carbon and nitrogen, and any emissions are not additional to pre-existing, natural baselines.[6]

If supported to do so, pastoralists are well suited to thrive under conditions of climate variability (Scoones, 1995; Davies and Nori, 2008; Goldman and Riosema, 2013; Pollini and Galaty, 2021; Rodgers, 2022; Marty et al., 2022), as they can generate livelihoods and produce high-quality protein in areas where other forms of agriculture are not feasible. This is especially important for poorer people, pregnant or lactating women, and young children. Arguing for reductions in livestock in pastoral areas in favour of arable agriculture or 'rewilding', including the mass planting of trees, may undermine livelihoods and would do little for the environment. A much more sophisticated, evidence-based approach is required, one that respects the diversity of livestock production, environmental dynamics, and dietary and livelihood needs. Simplistic interventions can do more harm than good. It is not the product (milk, meat, etc.) but the process of production and its material conditions of labour, capital, and environmental relations that matters (Houzer and Scoones, 2021). As much research now shows, low-impact pastoral systems can be good for livelihoods, the environment, and the climate (Köhler-Rollefson, 2023).

Mobility

People who move are often seen as a problem by those in charge. They cannot be controlled and are hardly ever taxed; their ways of life seem alien to settled populations and to state structures. This applies to pastoralists as it does to Travellers, Roma Gypsies, and others.[7] Movement is seen as disruptive to an ordered approach to development, and solutions often centre on highly regulated systems of movement corridors and designated areas for temporary settlement (FAO, 2022). Yet, as discussed at length in Chapter 3, flexible mobility is central to pastoral production systems, as well as pastoralists' connections with landscapes, their sense of identity, and their relationship to others.

Movement has an emotional, experiential dimension and is vital for confronting uncertainty and making use of variability. As we have discussed, pastoralism relies on the productive use of geographical and seasonal variability to maximize the opportunistic use of dispersed, often poor quality, grasslands. This involves the deployment of skilled labour to scout the environment and move animals, as well as negotiating access to resources. While boosting production through enhancing the nutrition of animals, such mobile practices can help offset disasters (such as drought, diseases, or conflict) by being flexible and so increasing reliability. Thus, movement cannot be just along strictly defined pathways but has to include an element of opportunism and responsiveness.

Too often, the mantra of increasing productivity through intensification is focused on a standard vision of settled forms of agriculture – by stabilizing production inputs and boosting output through irrigation, mechanization, and so on. But this model of intensification makes no sense in highly variable rangelands. In fact, it would have the opposite effect – settling down, while it may have benefits in terms of education, health care, and so on, will reduce productivity and undermine pastoralism. A 'sedentist' bias runs deep in policy thinking, rooted in a long history of what a 'modern', 'civilized' life should be, and reinforced by the agrarian and urban-focused thinking that dominates most policy arenas and methodologies (Pappagallo and Semplici, 2020).[8]

Such framings are difficult to shift, but a simple reflection highlights how important mobility of different forms is to contemporary life. People these days are increasingly mobile and connected; they move for work and leisure and rarely live in one place for their whole lives. Migration is central to the networked global economy, with people moving from country to country, place to place, to seek new opportunities – and so facilitating flows of information, goods, finances, and so on. In many ways, migrants must follow a lot of the same practices as pastoralists (Maru et al., 2022).

And, indeed, as Chapter 8 shows, migration and pastoralism are intimately connected through flows of people, finance, and information, in a flexible, 'liquid' form of modernity (Bauman, 2007). Appreciating this requires a rethink of the outmoded and inappropriate sedentary visions that dominate policy, recognizing the possibilities of mobility as the basis for production and a way of life. Pastoral policies need to facilitate mobility, managing the challenges – including conflicts between different land users – in the process (Davies et al., 2018).

Conflict

Pastoralists are often seen by outsiders as the source of conflict. The mythical images of the pastoral warrior claiming victory in raids over other groups or being a source of terror for settled populations – as the Huns or the Mongols were in the past – are difficult to dispel. Cross-border mobility makes pastoralists difficult to control by national state structures and laws. The response from centralized states, where political power is centred on the settled, agrarian heartlands, is typically a stabilization, securitization approach that aims to control populations by settling them (Bocco, 2006; Babalola and Onapajo, 2018), reducing the risks of secessionist insurgencies from the margins. A project of suppression alongside incorporation has long been part of state policy in pastoral areas.

A particular focus of much media and policy attention is herder–farmer conflicts, but herders and farmers have long had peaceful, symbiotic relations as manure, labour, and fodder are exchanged (Moritz, 2010; Krätli and Toulmin 2020). So what changes result in new conflicts? Conflicts may emerge because of increasing pressure on resources, as land is grabbed and removed from pastoral and small-scale agricultural use. This means that flexibility is removed from the system and tensions rise. Conflicts may arise because the nature of both farmers and herders has changed. With many large herds held by absentee owners, and managed by hired labourers, the intimate relationships that are central to managing resources jointly may be upset, while small-scale farmers may increasingly rely on chemical fertilizers rather than manure and so are less dependent on pastoralist inputs. In the past, pastoralist raiding was seen as a functionalist adaptation to localized resource scarcity, allowing restocking after a drought, but today, with an increasing prevalence of small arms, raiding is more wrapped up in local political contests (McCabe, 2010).

In other words, conflicts intensify when the social and political relations of production change, and this is the case across pastoral areas. This is added to by the dynamic of investment on the pastoral frontiers, with various forms of land, green, and water grabbing, alongside economic speculation, adding to resource competition and social tensions, while also providing the basis for the assertion of authority by pastoral elites (Fairhead et al., 2012; Lind et al., 2020a). Elite pastoralists may have strong alliances with the state, but others may feel abandoned and left behind by development efforts. The neglect by the state and the fragmentation of authority, combined with a waning of a sense of pastoral territorial identity as societies become increasingly unequal, results in a very different political dynamic in the rangelands (Korf et al., 2015). These tensions and the related grievances can be easily manipulated by factional politicians or by insurgent groups with their own interests and agendas. Pastoral youth can in turn be easily attracted by the lure of the weapons, wages, and power that smuggling networks or externally funded radical militias may offer (UNECA, 2017; Nori, 2022d).

Rather than seeing conflicts arising from the way of life of pastoralists on the periphery, the complex causes of conflicts have to be better understood. The solution is not to impose strategies of pacification, securitization, and control through a combination of military and developmental interventions. It is to build peace and security through an understanding of the often very long-term origins of conflicts, addressing their root causes, and drawing on 'vernacular' forms of peacebuilding that are compatible with changing pastoral livelihoods, and adapt institutions accordingly (Luckham, 2018; Lind et al., 2020b). Pastoralists are in fact the best allies to manage and create peace in vast dryland territories, but they have to be included in national and regional socio-economic and political agendas.[9]

Markets

A standard critique of pastoralists is that they are resistant to markets, that they have a 'cattle complex', and irrationally hold on to animals despite their market value. The response is the investment in various forms of market infrastructure – building marketplaces, roads, veterinary facilities, abattoirs, holding pens, and so on. Too often, such interventions fail; not because pastoralists are not market oriented but because the understanding of the pastoral market economy is inadequate (Mahmoud, 2008; Ng'asike et al., 2021). Pastoralists are of course deeply embedded in markets and have been for centuries. Look at the long-distance caravan trade in products across the Sahara or the focus on wool production and trade in Europe in the Middle Ages. Highly organized transnational trade networks also exist today: for example, Sahelian countries supply animal products to coastal countries to the south and Horn of Africa countries export millions of head to the Arabian Gulf each year. These generate very significant revenues for pastoral populations and for national states through taxation, while also serving the dietary needs of urban consumers, as well as supporting the regional integration of economic infrastructure (Catley and Aklilu, 2012).

As the middle class expands and urban centres boom, the demand for livestock products grows, and this results in the expansion of market networks, connecting producers on the margins to more established hubs where aggregation and bulk transport can be organized. Their operations have improved with the expansion of transport infrastructure (small trucks, motorbikes, and roads) and communication facilities (mobile phones and payments), which allow an adaptable, flexible form of marketing that responds to variability. With mobile herds and flocks, a single marketplace fixed in a certain site may not be appropriate, and instead, pastoralists sell in a network of more informal, temporary markets that pop up when the need arises.

Not all markets lead to a formal, terminal market (for example, the export of high-quality meat) and those with which pastoralists engage may be much more varied, with different products sold to different people across seasons (Roba et al., 2017). Camel-milk marketing in sub-Saharan Africa and the cheese trade in Sardinia (Chapter 5; Sadler et al., 2009) are good examples. They emerge through individual entrepreneurs or cooperative groups linking themselves to wider milk value chains, with benefits generated at all levels. These are 'real markets', not the result of abstract balances of supply and demand, but spaces where values are created through social relations and market connections embedded in social and cultural institutions. Understanding such pastoral markets is essential if pastoralists are to be supported in increasing income from their products. It is not that pastoralists are resistant to markets, but they often reject market interventions that are inappropriate.

Investments

Pastoral areas are often seen as the last frontier for investment. They are constructed as 'backward', empty, and remote, in need of 'development' (White et al., 2012). As a result, there are huge investments in irrigated agriculture, road infrastructure, urban developments, alternative solar and wind energy, tourism facilities, and conservation areas across these regions. The argument runs that such investments will reduce poverty in these areas, encourage pastoralists to settle, offset the problems of depopulation and emigration, and provide greater contributions to the national economy, while also improving the management of natural resources. As these areas have long been neglected by the state, private investors, and other development efforts, there is some value in such arguments. But the question remains whether these are the right sort of investments, compatible with providing adequate support to pastoral production and livelihoods.

As discussed above, it is not that pastoral production has no value – indeed, in some economies, it is a very significant and increasing part of the national food security and (export) economy (Hesse and MacGregor, 2009). As pastoralism diversifies and differentiates, some pastoralists follow a trajectory of animal commercialization, while others drop out of direct production of animals but support a service economy linked to pastoralism (transport, veterinary care, fodder supply, etc.), while still others get involved in processing (producing cheese and other milk or meat products, hides and skins, for example) (Catley et al., 2012). Such diversification has resulted in the growth of small towns in pastoral areas, which in turn attract people and businesses as part of a virtuous cycle of investment, growth, and improved service provision (Fratkin, 1997; Little, 2012). This dynamic is barely recognized and is poorly studied, as the assumptions that remain so prevalent are that pastoralism is dead and alternatives must be found.

External investments that are compatible with pastoralism can support the ability of pastoralists to live with and from variability. Some may have low land take and do not grab key resources (this includes some forms of energy investment but usually not irrigated farming); some may be able to benefit from working with pastoralists (in schemes where herders can become rangers and sustain conservation efforts alongside animal production); and some might provide support to existing activities to diversify and expand livelihood from livestock (such as processing and marketing animal products).

Experience attests that investing in transport and communication facilities and in provision of basic services (including animal health) can magnify socio-economic potentials in pastoral areas. But most investments, as currently conceived, are not thought about in this way. The dislocation of pastoralists through exclusionary forms of investment is growing – and some of this is designed as part of 'environmental' initiatives tagged as biodiversity conservation, green energy, or tree planting for carbon sequestration, as discussed above. As we have seen through the case studies, pastoralism can be part of a resilient, vibrant economy in marginal areas, but this requires a different approach to investment, which crucially means involving pastoral communities in identifying and designing investments suited to their own needs and contexts (Gomes, 2006; Krätli et al., 2013).

Disasters

Dryland pastoral areas are often portrayed as sites of recurrent disasters, where extreme poverty and destitution result from the failure of pastoralism to support a fast-growing population. The result is the disaster, relief, and dependency cycle, where former pastoralists continue to be reliant on humanitarian aid and where large refugee settlements are established (Catley and Cullis, 2012; Catley, 2017). The alternative, it is argued, must be migration away from drylands to other more productive areas, or investment in alternative, non-pastoral livelihoods. Such a narrative, however, lacks a complex understanding of how disasters strike and how they can be averted. The huge investments in early warning systems, cash transfers, insurance schemes, assisted migration, and alternative livelihood investments are often misplaced. Too often, such programmes assume that the problem is a singular disaster (such as drought) affecting an area uniformly and that better information (through early warning) and effective finance (via insurance, for example) will offset a crisis. But disasters emerge through the convergence of multiple factors, acting concurrently and with cascading effects, and so have deep uncertainties. They cannot be predicted, anticipated, and planned for in the standard way that current systems assume. Instead, a disaster and humanitarian approach must take account of uncertainty (and ignorance) and build on pastoralists' own capacities as 'high-reliability professionals', accepting that informal redistributions through social institutions are effective (Caravani et al., 2022; Lind et al., 2022).

Humanitarian assistance in the form of transfers of food, cash, or assets is often based on 'targeting' what are assumed to be fixed, settled households organized as individual entities. The collective forms of solidarity and networked arrangements that are a central part of the local moral economy are often overlooked. As Chapter 6 describes, among the Boran pastoralists of Isiolo in northern Kenya, intersecting forms of moral economy allow for collective support and redistribution when disasters strike. This includes the ability to restock animals after loss, support those in need of labour, and the provision of food and income. All of these involve collective forms of organization and solidarity, with connections across family and clan vitally important. As Chapter 7 explains, insurance schemes are increasingly seen as part of the social protection portfolio, but a market-based instrument focused on the individualized ownership of animals and targeted at offsetting the risks of a single peril – drought – may not work across all forms of uncertainty and must be combined with other forms of coping and reliability management.

This is not an argument for halting humanitarian assistance, social protection, and livelihood support. Instead, it is one for making such approaches more attuned to pastoral settings, with diversification approaches compatible with the support of the wider pastoral economy, including opportunities to return to pastoralism later.[10] Increasing resilience in the face of recurrent crises requires generating reliability, and this means supporting 'high-reliability' pastoralists – as individuals who are embedded in networks – to scan the horizon for future threats as well as opportunities; to network amongst each other and with others, including aid agencies and the state; to facilitate support; and to develop a flexible system of response that is not reliant on standardized risk assessment approaches (Roe, 2020; Konaka and Little, 2021; Tasker and Scoones, 2022). Some responses may indeed mean moving away from pastoralism and seeking alternative livelihoods, but for many, such alternatives are viewed as only temporary, a route to getting back into pastoralism. Having the flexibility to move in and out of pastoralism is therefore important as a long-term response to variable, disaster-prone conditions (Catley, 2017).

Table 9.1 offers a very brief and necessarily grossly simplified summary of the shifts in narratives that are required to reframe pastoral policy so that it is more attuned to variability and uncertainty across the seven themes just discussed. This suggests a new agenda for policy advocacy and research in pastoral areas, where uncertainty is central, seeing a shift from control, planning, and directive management to a more flexible, responsive, caring approach (Scoones and Stirling 2020).

Table 9.1 Reframing pastoral policy

Theme	Existing narrative	New narrative
Environmental degradation	Rangelands are degraded through overgrazing resulting in desertification; they are in need of restoration, such as through destocking and tree planting.	Rangelands are open, non-equilibrium systems where variability is inherent and mixed grass–tree combinations are maintained by grazing and fire. Pastoralism supports diverse ecosystem functions, including biodiversity protection.
Climate change	Livestock-based greenhouse emissions cause climate change and livestock production should be curtailed through diet change and/or shifts to other forms of protein production.	Different livestock systems produce different impacts. It is not the product but the process that matters. In contrast to industrial contained livestock systems, pastoralism is a low-impact system with potentials for carbon/nitrogen sequestration.
Mobility	Mobility is disruptive, 'backward', and unproductive; pastoralists should be settled and become 'modern', adopting a sedentary, civilized way of life.	Mobility is essential for making productive use of variable resources, whether through movement of grazing animals or people migrating. Flexible mobility of people and animals should be facilitated, not curtailed.
Conflict	Conflicts caused by pastoralists trigger instability and violence. Pastoral areas need to be stabilized and pacified through militarized security interventions.	Conflicts are the result of long-term neglect of pastoral areas by the state, the fragmentation of local authority, increasing social stratification, and resource grabbing, exacerbated by the instrumentalization of grievances and the availability of small arms.
Markets	Pastoralists are resistant to market-based development. They need to be incorporated into modern, market systems through new investments.	Pastoralists have long been engaged with markets and trade, and continue to be so through diverse, networked, nested arrangements compatible with pastoral production systems.
Investment	Pastoral areas are remote and in need of investment. Investments will have trickle-down growth effects on the economy, encouraging pastoralists to settle, modernize, and move away from pastoralism.	Investment to support pastoral production is much needed, but it should invest in pastoralism's future and enhance its reliability, including diversifying and expanding the pastoral economy.
Disasters	Recurrent disasters plague pastoral areas, resulting in long-term reliance on aid. Early warning systems and disaster risk management systems, including insurance, are required alongside encouraging alternative livelihoods outside pastoralism.	Pastoralists have many existing responses to variability, including sudden shocks. These are embedded in social networks and rely on caring moral economy practices, which respond to unfolding uncertainties. Support for high-reliability professionals and their networks should be central to building local resilience and more effective responsiveness.

An agenda for policy action

What then would a policy agenda that is centred on 'seeing like a pastoralist' rather than a state planner or policymaker – or even an aid or humanitarian agency worker or an urban environmentalist – look like? In the above sections, and indeed throughout the book, we have outlined some of the changes that are needed. Accepting that variability – and therefore uncertainty – is central to pastoral settings is the starting point. Variability is a source of production and an opportunity, not something to be suppressed, controlled, and eliminated. Instead, support for mechanisms to generate reliability, and therefore resilience, in response to high levels of variability is required. This requires very different ways of thinking, which contrast with the standard views of modernization and development centred on control and stability.

Such an alternative narrative about pastoral development, in turn, means a fundamental recasting of debates about environmental degradation and restoration, around climate change and biodiversity, and around development interventions in support of market integration, infrastructure investment, migration governance, disaster risk management, and humanitarian assistance. As the discussions in this book have shown, the consequences are far-reaching, with major changes required in the way people are trained, institutions are designed, policies are made, and funds are deployed. This will not happen overnight, as the existing biases are so entrenched.

However, rethinking some approaches through pilot efforts – for example, supporting pastoralists as reliability professionals in social protection and humanitarian assistance programmes, or redesigning resilience and livelihood interventions with variability and reliability in mind – could begin to make in-roads into constructing practical alternatives. In the same way, decision-making around the global debates concerning climate and biodiversity needs to be recast so as to avoid a simplistic one-size-fits-all anti-livestock narrative. Equally, as plans for investment in drylands areas are developed, these must assess the compatibility with pastoral systems, without assuming such areas are empty and unused, and recognize that there are no better alternatives in such areas in terms of food production, environmental management, and territorial control than pastoralism.

None of these urgently needed changes will happen unless pastoralists are involved in such debates, with the confidence to articulate an alternative view and confront the structural, political-economic biases that limit pastoralists' room for manoeuvre. Pastoralists' voices in international debates are important, but too often elite pastoralists on international platforms are co-opted into mainstream positions. Developing grounded, evidence-based alternative narratives to support pastoral organizations to mobilize around an alternative view is a central task, and we see this book as a very limited part of this struggle.[11] Supporting forms of solidarity and organizational capacity among pastoralists is challenging, as pastoralists are dispersed and highly differentiated. Avoiding the pigeonholing of pastoralists in a romantic construction of an 'indigenous' people is important: pastoralism, as we have explained, is a highly modern form of production. The United Nations International Year for Rangelands and Pastoralism, which has been announced for 2026,[12] will be an important moment for such mobilization of both ideas and action, and this book hopefully provides some empirical foundations as well as conceptual support for such efforts.

As we have hinted at throughout this book, the argument for pastoralism must be made on its own terms – pastoralism is a low-impact, productive system supporting marginal livelihoods in often challenging areas where alternative forms of use are limited – but there is also a bigger argument for taking pastoralism seriously. And this is the very basic connection between uncertainty and development highlighted in Chapter 1. As shown, pastoralists are well practised at responding to uncertainty, in converting high variability into reliable services, and therefore livelihoods, in harsh settings.

Such conditions of uncertainty are faced by many people across the world today; this is not just a challenge for those living in the periphery, on the margins. If we are to respond to climate change, market volatility, changing environments, migratory flows, more frequent pandemics, and rising conflict, we can and must learn from those who have developed the capacities to live with and from uncertainty (Scoones, 2019, 2022c). Thinking about how pastoralists respond to uncertainty can be important, whether thinking about pastoral mobility when constructing human migration policies (Maru et al., 2022); designing social assistance and humanitarian relief approaches that avoid centralized risk-based approaches (Caravani et al., 2022); fostering market integration dynamics that build around local practices and networks (Nori, 2023); supporting knowledge networking and exchange as part of extension efforts to increase reliability (Tasker and Scoones, 2022); redesigning insurance schemes to support a more varied response (Johnson et al., 2023); thinking about preparedness for pandemics (Leach et al., 2022) or disasters more generally (Srivastava and Scoones, forthcoming); or even rethinking banking, finance, and economic policymaking itself (DeMartino et al., forthcoming; Scoones, 2020).

In our turbulent world, where uncertainties affect us all, insights from pastoralism can be enormously helpful. Perhaps, above all, this is why linking pastoralism, uncertainty, and development is such a central challenge for our times.

Notes

1. We have not looked at the Americas, Australia, or southern Africa where different policy regimes exist and different policy dynamics occur, but similar narratives framing policy emerge (see Scoones, 2021).
2. See, for example, a review of the East African context and associated policy challenges (Muhereza, 2017).

3. See https://au.int/en/documents/20110131/policy-framework-pastoralism-africa-securing-protecting-and-improving-lives; https://icpald.org/wp-content/uploads/2021/05/Legal-Policy-and-Institutional-Frameworks-in-IGAD-Region.pdf.
4. Such protocols were first implemented in West Africa (https://ecpf.ecowas.int/wp-content/uploads/2016/01/CrossBorder-Transhumance-WA-Final-Report-1.pdf) and have more recently been agreed in Eastern Africa (http://www.celep.info/wp-content/uploads/2020/12/2020-IGAD-protocol-on-transhumance-final-endorsed-version.pdf).
5. See www.pastres.org/biodiversity
6. https://pastres.org/livestock-report/; Manzano and White (2019).
7. https://pastres.org/2022/02/25/the-last-nomads-proposed-new-law-undermine-gypsy-traveller-communities-nomadic-lifestyles-uk/
8. https://pastres.org/2022/03/18/how-sedentist-approaches-to-land-and-conservation-threaten-pastoralists/
9. As with the governmental declarations issued in 2013 in N´Djamena and Nouakchott (http://www.pasto-secu-ndjamena.org/classified/N_Djamena_Declaration_eng.pdf; https://rr-africa.woah.org/wp-content/uploads/2000/11/nouakchott-1.pdf).
10. See https://au.int/en/documents/20110131/policy-framework-pastoralism-africa-securing-protecting-and-improving-livesSee the Livestock Emergency Guidelines and Standards initiative, https://www.livestock-emergency.net/
11. See also the Pastoralism primer produced in collaboration with the Transnational Institute and available in multiple languages, https://www.tni.org/en/publication/livestock-climate-and-the-politics-of-resources.
12. https://www.iyrp.info/

References

Babalola, D. and Onapajo, H. (eds) (2018) *Nigeria, a Country under Siege: Issues of Conflict and its Management in Democratic Nigeria*, Cambridge Scholars Publishing, Newcastle upon Tyne.

Bauman, Z. (2007) *Liquid Times: Living in an Age of Uncertainty*, Polity Press, Cambridge.

Behnke, R.H., Scoones, I. and Kerven, C. (eds) (1993) *Range Ecology at Disequilibrium: New Models of Natural Variability and Pastoral Adaptation in African Savannas*, Overseas Development Institute, London.

Benjaminsen, Tor A. and Ba, B. (2019) 'Why do pastoralists in Mali join jihadist groups? A political ecological explanation', *Journal of Peasant Studies* 46: 1–20 <https://doi.org/10.1080/03066150.2018.1474457>.

Bocco, R. (2006) 'The settlement of pastoral nomads in the Arab Middle East: international organizations and trends in development policies, 1950–1990', in D. Chatty (ed.), *Nomadic Societies in the Middle East and North Africa*, Handbook of Oriental Studies, Section 1: The Near and Middle East, vol 81, pp. 302–32, Brill, Leiden.

Bond, W.J. (2019) *Open Ecosystems: Ecology and Evolution Beyond the Forest Edge*, Oxford University Press, Oxford <https://doi.org/10.1093/oso/9780198812456.001.0001>.

Bond, W.J., Stevens, N., Midgley, G.F. and Lehmann, C.E. (2019) 'The trouble with trees: afforestation plans for Africa', *Trends in Ecology & Evolution* 34: 963–65 <https://doi.org/10.1016/j.tree.2019.08.003>.

Briske, D.D., Fuhlendorf, S.D. and Smeins, F.E. (2003) 'Vegetation dynamics on rangelands: a critique of the current paradigms', *Journal of Applied Ecology* 40: 601–14 <https://doi.org/10.1046/j.1365-2664.2003.00837.x>.

Caravani, M., Lind, J., Sabates-Wheeler, R. and Scoones, I. (2022) 'Providing social assistance and humanitarian relief: the case for embracing uncertainty', *Development Policy Review* 40: e12613 <https://doi.org/10.1111/dpr.12613>.

Catley, A. (2017) *Pathways to Resilience in Pastoralist Areas: A Synthesis of Research in the Horn of Africa*, Feinstein International Center, Tufts University, Boston MA.

Catley, A. and Aklilu, Y. (2012) 'Moving up or moving out? Commercialization, growth and destitution in pastoralist areas', in A. Catley, J. Lind and I. Scoones (eds), *Pastoralism and Development in Africa: Dynamic Change at the Margins*, pp. 85–97, Routledge, London <https://doi.org/10.4324/9780203105979>.

Catley, A. and Cullis, A. (2012) 'Money to burn? Comparing the costs and benefits of drought responses in pastoralist areas of Ethiopia', *Journal of Humanitarian Assistance* 1548: 1–8.

Catley, A., Lind, J. and Scoones, I. (eds) (2012) *Pastoralism and Development in Africa: Dynamic Change at the Margins*, Routledge, London <https://doi.org/10.4324/9780203105979>.

Davies, J. and Nori, M. (2008) 'Managing and mitigating climate change through pastoralism', *Policy Matters* 16: 127–62.

Davies, J., Olgali, C., Slobodian, L., Roba, G. and Ouedraogo, R. (2018) *Crossing Boundaries: Legal and Policy Arrangements for Cross-Border Pastoralism*, in G. Velasco-Gil and N. Maru (eds), FAO and IUCN, Rome and Gland.

DeMartino, G., Grabel, I. and Scoones, I. (forthcoming) 'Economics for an uncertain world'.

Fairhead, J., Leach, M. and Scoones, I. (2012) 'Green grabbing: a new appropriation of nature?', *Journal of Peasant Studies* 39: 237–61 <https://doi.org/10.1080/03066150.2012.671770>.

FAO (2022) *Making Way: Developing National Legal and Policy Frameworks for Pastoral Mobility*, FAO Animal Production and Health Guidelines 28, FAO, Rome <https://doi.org/10.4060/cb8461en>.

Fratkin, E. (1997) 'Pastoralism: governance and development issues', *Annual Review of Anthropology* 26: 235–61 <https://www.jstor.org/stable/2952522>.

Gabbert, E.C., Gebresenbet, F., Galaty, J.G. and Schlee, G. (eds) (2021) *Lands of the Future: Anthropological Perspectives on Pastoralism, Land Deals and Tropes of Modernity in Eastern Africa*, Berghahn Books, New York NY.

Garcia, A.K., Haller, T., van Dijk, H., Samimi, C. and Warner, J. (eds) (2022) *Drylands Facing Change: Interventions, Investments and Identities*, Routledge, Abingdon.

Goldman, M. and Riosmena, F. (2013) 'Adaptive capacity in Tanzanian Maasailand: changing strategies to cope with drought in fragmented landscapes', *Global Environmental Change* 23: 588–97.

Gomes, N. (2006) *Access to Water, Pastoral Resource Management and Pastoralists' Livelihoods: Lessons Learned from Water Development in Selected Areas of Eastern Africa (Kenya, Ethiopia, Somalia)*, Food and Agriculture Organization, Nairobi.

Gongbuzeren, Huntsinger, L. and Li, W. (2018) 'Rebuilding pastoral social-ecological resilience on the Qinghai-Tibetan Plateau in response to changes in policy, economics, and climate', *Ecology and Society* 23: 21 <https://doi.org/10.5751/ES-10096-230221>.

Haan, C. de (1993) *An Overview of the World Bank's Involvement in Pastoral Development*, paper presented at the Donor Consultation Meeting on Pastoral National Resource Management and Pastoral Policies for Africa organized by the United Nations Sudano-Sahelian Office, Paris, December 1993.

Hesse, C. and MacGregor, J. (2009) *Arid Waste? Reassessing the Value of Dryland Pastoralism*, IIED briefing, International Institute for Environment and Development, London. Available from: <https://www.iied.org/17065iied> [accessed 17 January 2023].

Houzer, E. and Scoones, I. (2021) *Are Livestock Always Bad for the Planet? Rethinking the Protein Transition and Climate Change Debate*, PASTRES, Brighton <https://doi.org/10.19088/STEPS.2021.003>.

Johnson, L., Mohamed, T., Scoones, I. and Taye, M. (2023) 'Uncertainty in the drylands: rethinking in/formal insurance from pastoral East Africa', *Environment and Planning A*.

Kerven, C., Robinson, S. and Behnke, R. (2021) 'Pastoralism at scale on the Kazakh rangelands: from clans to workers to ranchers', *Frontiers: Sustainable Food Systems* 4: 590401 <https://doi.org/10.3389/fsufs.2020.590401>.

Köhler-Rollefson, I. (2023) *Hoofprints on the Land: How Traditional Herding and Grazing Can Restore the Soil and Bring Agriculture Back in Balance with the Earth*, Chelsea Green, London.

Konaka, S. and Little, P. (2021) 'Introduction: rethinking resilience in the context of East African pastoralism', *Nomadic Peoples* 25: 165–80.

Korf, B., Hagmann, T. and Emmenegger, R. (2015) 'Re-spacing African drylands: territorialization, sedentarization and indigenous commodification in the Ethiopian pastoral frontier', *Journal of Peasant Studies* 42: 881–901 <https://doi.org/10.1080/03066150.2015.1006628>.

Krätli, S. and Toulmin, C. (2020) *Farmer-Herder Conflict in sub-Saharan Africa?* International Institute for Environment and Development (IIED), London.

Krätli, S., Huelsebusch, C., Brooks, S. and Kaufmann, B. (2013) 'Pastoralism: a critical asset for food security under global climate change', *Animal Frontiers* 3: 42–50 <https://doi.org/10.2527/af.2013-0007>.

Leach, M., MacGregor, H., Ripoll, S., Scoones, I. and Wilkinson, A. (2022) 'Rethinking disease preparedness: incertitude and the politics of knowledge', *Critical Public Health* 32: 82–96 <https://doi.org/10.1080/09581596.2021.1885628>.

Lind, J., Okenwa, D. and Scoones, I. (eds) (2020a) 'The politics of land, resources and investment in Eastern Africa's pastoral drylands', in J. Lind, D. Okenwa and I. Scoones (eds.), *Land Investment and Politics: Reconfiguring Eastern Africa's Pastoral Drylands*, pp. 1–32, James Currey, Woodbridge.

Lind, J., Sabates-Wheeler, R., Caravani, M., Biong Deng Kuol, L. and Manzolillo Nightingale, D. (2020b) 'Newly evolving pastoral and post-pastoral

rangelands of Eastern Africa', *Pastoralism* 10: 1–14 <https://doi.org/10.1186/s13570-020-00179-w>.
Lind, J., Sabates-Wheeler, R., Hoddinott, J. and Taffesse, A.S. (2022) 'Targeting social transfers in Ethiopia's agro-pastoralist and pastoralist societies', *Development and Change* 53: 279–307 <https://doi.org/10.1111/dech.12694>.
Little, P. (2012) 'Reflections on the future of pastoralism in the Horn of Africa', in A. Catley, J. Lind and I. Scoones (eds), *Pastoralism and Development in Africa: Dynamic Change at the Margins*, pp. 243–49, Routledge, London.
Luckham, R. (2018) 'Building inclusive peace and security in times of unequal development and rising violence', *Peacebuilding* 6: 87–110 <https://doi.org/10.1080/21647259.2018.1449185>.
Mahmoud, H. (2008) 'Risky trade, resilient traders: trust and livestock marketing in northern Kenya', *Africa* 78: 561–81 <https://doi.org/10.3366/E0001972008000442>.
Manzano, P. and White, S.R. (2019) 'Intensifying pastoralism may not reduce greenhouse gas emissions: wildlife-dominated landscape scenarios as a baseline in life-cycle analysis', *Climate Research* 77: 91–7 <https://doi.org/10.3354/cr01555>.
Marty, E., Bullock, R., Cashmore, M., Crane, T. and Eriksen, S. (2022) 'Adapting to climate change among transitioning Maasai pastoralists in southern Kenya: an intersectional analysis of differentiated abilities to benefit from diversification processes', *Journal of Peasant Studies* <https://doi.org/10.1080/03066150.2022.2121918>.
Maru, N. (2022) '*Haal, haal ne haal* [to Walk, Walk and Walk]: Exploring the Temporal Experiences of Pastoral Mobility in Western India', doctoral dissertation, University of Sussex.
Maru, N., Nori, M., Scoones, I., Semplici, G. and Triandafyllidou, A. (2022) 'Embracing uncertainty: rethinking migration policy through pastoralists' experiences', *Comparative Migration Studies* 10: 1–18 <https://doi.org/10.1186/s40878-022-00277-1>.
McCabe, J.T. (2010) *Cattle Bring us to our Enemies: Turkana Ecology, Politics, and Raiding in a Disequilibrium System*, University of Michigan Press, Ann Arbor MI.
Mohamed, T. (2022) 'The Role of the Moral Economy in Response to Uncertainty among the Pastoralists in Northern Kenya', doctoral dissertation, University of Sussex. <https://sro.sussex.ac.uk/id/eprint/110472/>
Moritz, M. (2010) 'Understanding herder–farmer conflicts in West Africa: outline of a processual approach', *Human Organization* 69: 138–48 <http://www.jstor.org/stable/44148597>.
Muhereza, E.F. (2017) *Pastoralist and Livestock Development in Karamoja, Uganda: A Rapid Review of African Regional Policy and Programming Initiatives*. Karamoja Resilience Support Unit, USAID/Uganda, Kampala. Available from: <https://karamojaresilience.org/wp-content/uploads/2021/05/tufts_1742_krsu_review_policies_karamoja_v3_online.pdf>.
Ng'asike, O.A.P., Hagmann, T. and Wasonga, O.V. (2021) 'Brokerage in the borderlands: the political economy of livestock intermediaries in northern Kenya', *Journal of Eastern African Studies* 15: 168–88 <https://doi.org/10.1080/17531055.2020.1845041>.

Nori, M. (2019) *Herding through Uncertainties – Principles and Practices: Exploring the Interfaces between Pastoralists and Uncertainty: Results from a Literature Review*, Working Paper 69, Global Governance Programme, EUI Robert Schuman Centre, Firenze. Available from: <https://cadmus.eui.eu/handle/1814/64228>.

Nori, M. (2022a) *Assessing the Policy Frame in Pastoral Areas of Asia*, Policy Paper 2022/03, Global Governance Programme, EUI Robert Schuman Centre, Firenze. Available from: <https://cadmus.eui.eu/handle/1814/74316>.

Nori, M. (2022b) *Assessing the Policy Frame in Pastoral Areas of Europe*, Policy Paper 73811, Global Governance Programme, EUI Robert Schuman Centre, Firenze. Available from: <https://hdl.handle.net/1814/73811>.

Nori, M. (2022c) *Assessing the Policy Frame in Pastoral Areas of West Asia and North Africa*, Policy Paper 74315, Global Governance Programme, EUI Robert Schuman Centre, Firenze. Available from: <http://hdl.handle.net/1814/74315>.

Nori, M. (2022d) *Assessing the Policy Frame in Pastoral Areas of Sub-Saharan Africa*, Policy Paper 74314, Global Governance Programme, EUI Robert Schuman Centre, Firenze. Available from: <http://hdl.handle.net/1814/74314>.

Nori, M. (2023) High quality, high reliability: The dynamics of camel milk marketing in northern Kenya. Pastoralism 13 (9), <https://doi.org/10.1186/s13570-022-00265-1>.

Pappagallo, L. (2022) '*"Partir Pour Rester?"*: To Leave in Order to Stay? The Role of Absence and Institutions in Accumulation by Pastoralists in Southern Tunisia', doctoral thesis, Sussex University <https://sro.sussex.ac.uk/id/eprint/109122/>.

Pappagallo, L. and Semplici, G. (2020) 'Editorial introduction. Methodological mess: doing research in contexts of high variability', *Nomadic Peoples* 24: 179–94 <https://doi.org/10.3197/np.2020.240201>.

Pollini, J. and Galaty, J.G. (2021) 'Resilience through adaptation: innovations in Maasai livelihood strategies', *Nomadic Peoples* 25: 278–311 <https://doi.org/10.3197/np.2021.250206>.

Roba, G.M., Lelea, M.A. and Kaufmann, B. (2017) 'Manoeuvring through difficult terrain: how local traders link pastoralists to markets', *Journal of Rural Studies* 54: 85–97 <https://doi.org/10.1016/j.jrurstud.2017.05.016>.

Robinson, L. and Flintan, F. (2022) 'Can formalisation of pastoral land tenure overcome its paradoxes? Reflections from East Africa', *Pastoralism* 12: 1–12 <https://doi.org/10.1186/s13570-022-00250-8>.

Rodgers, C. (2022) *Equipped to Adapt? A Review of Climate Hazards and Pastoralists' Responses in the IGAD Region*, IOM and ICPALD, Nairobi.

Roe, E. (2020) *A New Policy Narrative for Pastoralism? Pastoralists as Reliability Professionals and Pastoralist Systems as Infrastructure*, STEPS Working Paper 113, STEPS Centre, Brighton. Available from: <https://opendocs.ids.ac.uk/opendocs/bitstream/handle/20.500.12413/14978/STEPS_WP_113_Roe_FINAL.pdf?sequence=105&isAllowed=y>.

Sadler, K., Kerven, C., Calo, M., Manske, M. and Catley, A. (2009) *Milk Matters: A Literature Review of Pastoralist Nutrition and Programming Responses*, Feinstein International Center, Tufts University and Save the Children, Addis Ababa.

Scoones, I. (ed.) (1995) *Living with Uncertainty: New Directions in Pastoral Development in Africa*. Intermediate Technology Publications, Rugby. Available from: <https://practicalactionpublishing.com/book/1264/living-with-uncertainty>.

Scoones, I. (2019) 'What is uncertainty and why does it matter?', *STEPS Working Paper* 105, STEPS Centre, Brighton. Available from: <https://opendocs.ids.ac.uk/opendocs/bitstream/handle/20.500.12413/14470/STEPSWP5_Scoones_final.pdf?sequence=1&isAllowed=y>.

Scoones, I. (2020) 'What pastoralists know', *Aeon*. Available from: <https://aeon.co/essays/what-bankers-should-learn-from-the-traditions-of-pastoralism>.

Scoones, I. (2021) 'Pastoralists and peasants: perspectives on agrarian change', *Journal of Peasant Studies* 48: 1–47 <https://doi.org/10.1080/03066150.2020.1802249>.

Scoones, I. (2022a) 'A new politics of uncertainty: towards convivial development in Africa', in C. Greiner, S. van Wolputte and M. Bollig (eds), *African Futures*, pp. 101–10, Brill, Leiden <https://doi.org/10.1163/9789004471641_009>.

Scoones, I. (2022b) 'What is environmental degradation, what are its causes, and how to respond?' *IDS Working Paper* 577, IDS, Brighton. Available from: <https://opendocs.ids.ac.uk/opendocs/handle/20.500.12413/17608>.

Scoones, I. (2022c) *Livestock, Climate and the Politics of Resources: A Primer*, Transnational Institute, Amsterdam. Available from: <https://www.tni.org/en/publication/livestock-climate-and-the-politics-of-resources>.

Scoones, I. and Stirling, A. (eds) (2020) *The Politics of Uncertainty: Challenges of Transformation*, Routledge, London. Available from: <https://library.oapen.org/handle/20.500.12657/39938>.

Simula, G. (2022) 'Pastoralism 100 Ways: Navigating Different Market Arrangements in Sardinia', doctoral dissertation, University of Sussex. <https://sro.sussex.ac.uk/id/eprint/109485/>.

Srivastava, S. and Scoones, I. (forthcoming) 'Rethinking disaster response: building reliability under climatic uncertainty'.

Steinfeld, H., Gerber, P., Wassenaar, T., Castel, V., Rosales, M. and de Haan, C. (2006) *Livestock's Long Shadow: Environmental Issues and Options*, FAO, Rome. Available from: <https://www.fao.org/3/a0701e/a0701e00.htm>.

Swift, J. (1996) 'Desertification. Narratives, winners and losers', in M. Leach and R. Mearns (eds), *The Lie of the Land: Challenging Received Wisdom on the African Environment*, pp.73–90, James Currey, Oxford.

Tasker, A. and Scoones, I. (2022) 'High reliability knowledge networks: responding to animal diseases in a pastoral area of northern Kenya', *Journal of Development Studies* 58: 968–88 <https://doi.org/10.1080/00220388.2021.2013469>.

Taye, M. (2022) 'Financialisation of Risk among the Borana Pastoralists of Ethiopia: Practices of Integrating Livestock Insurance in Responding to Risk', PhD dissertation, Institute of Development Studies, University of Sussex, Brighton. Available from: <https://sro.sussex.ac.uk/id/eprint/109001/>.

Tsering, P. (2022) 'Institutional Hybridity: Rangeland Governance in Amdo, Tibet', doctoral dissertation, University of Sussex. Available from: <https://core.ac.uk/download/pdf/519857982.pdf>.

UNECA (2017) *New Fringe Pastoralism: Conflict and Insecurity and Development in the Horn of Africa and the Sahel*, UN Economic Commission for Africa, Addis Ababa. Available from: <https://hdl.handle.net/10855/23727>.

White, B., Borras Jr, S.M., Hall, R., Scoones, I. and Wolford, W. (2012) 'The new enclosures: critical perspectives on corporate land deals', *Journal of Peasant Studies* 39: 619–47 <https://doi.org/10.1080/03066150.2012.691879>.

Yılmaz, E., Zogib, L., Urivelarrea, P. and Demirbaş Çağlayan, S. (2019) 'Mobile pastoralism and protected areas: conflict, collaboration and connectivity', *Parks* 25: 7–24 <https://doi.org/10.2305/IUCN.CH.2019.PARKS-25-1EY.en>.

Index

agriculture
  diversification 67
  European Common Agricultural Policy (CAP) 67, 121
  expanding 101
  genetically modified crops 40, 42, 76–7
  migration for work 112
  and pastoralism 15
Amdo Tibet, China 10–11, 22–3, 26, 51
  adaptation in Amdo Tibet context 55–6
  Amdo Tibet map 52
  lake expansion 28–9, 53–4, 55–6, 59–60
  negotiating solutions for adaptive responses 60–1
  pluralist resource governance 59–60
  role of monasteries in resource governance 57–9
  seeing and concepts of uncertainty 52–4
  state governance and hybrid arrangements 56–7
animal feed/nutrition 68, 71–2, 75, 76–7
  insured and uninsured families 99
assemblage practices and processes 57, 60, 61

'backward' narrative 6, 41, 67, 69, 89, 123, 128
Belt and Road Initiative 10, 120
biodiversity crisis *see* environmental degradation
Borana, southern Ethiopia 13–4, 95–6
  insurance and its assumptions 94–5
  insurance and local responses, combining 98–102

  insured and uninsured pastoralists, comparing 96–8
  locust plague 27–8
Buddhism
  concept of change in motion 53–4
  role of monasteries 57–9

camel meat and milk production 85–6, 89–90
cheese making *see* Sardinia, Italy
China and Central Asia: pastoral policies 120
class divisions *see* wealth differences
climate change 1, 3, 6, 12, 16, 29, 35, 41, 51–60, 79, 93, 124–5, 131–3
  extreme weather events 66
  lake expansion 28–9, 53–4, 55–6, 59–60
  nature-positive contribution of pastoralism 6–7
collaboration
  identities, relationships, and 88–9
  negotiating solutions for adaptive responses 60–1
  resource pooling 85, 112–6
collective herding arrangement *see* Douiret, southern Tunisia
collective vs individual responses 101, 102, 116
colonial perspective 6, 7
common, commoning etc. 62, 67, 99–101
comradeship and resource pooling 85
conflict 122, 126–7
cooperative dairies 66, 74–5
Covid-19 pandemic 23–4, 28, 89–90
critical infrastructure 3–4
crops *see* agriculture

development
  challenges for 8–9
  efficiency and productivity narrative 68
  and humanitarian support 87–8, 89
  Kachchh, Gujarat, India 42–4, 47–8
  local conditions and external interventions 4, 86–90
  reframing 2–4
  state policies 56–7
  World Bank report 119
  see also pastoral policies
disasters 129–30
diversification 15, 67, 75, 85–6, 89–90
documentary photo/video 26–7
Douiret, southern Tunisia 14, 27, 31–2, 107–9
  conceptions and management of uncertainty 31–2
  evolution of informal collective resource management 115–6
Douiret map *108*
  khlata (collective herding arrangement) 14, 109, 112–7
  migration and 'absence' 32, 110–2
  pooling resources and collective arrangements 112–5
droughts 5, 8, 21, 66, 31, 79, 83–5, 93–5, 101–3, 113–4, 122–6
  livestock insurance *see* Borana, southern Ethiopia

elders 24, 30, 55, 112
environmental degradation 7–8, 123–4
  grazing ban 59, 60
European Union 67, 121
extreme weather events 66
  see also droughts; hailstorm

Facebook 23–4, 32
fire
  and extreme weather events 66
  nature-positive contribution of 7–8

food consumption, changing 99–100
food crises, Kenya 83

gender 13, 15, 22, 23, 103, 116, 121
  insurance 96, 98
  migration and absence 111–2
  see also networks
genetically modified crops 40, 42, 76–7
governance 4, 16, 22, 120, 132
  resource 2, 57–9
  see also hybrid rangeland governance
grassland 5, 28, 39, 55, 101, 123, 131
grazing 5, 10–4, 30–1, 44, 46, 77, 83, 109
  ban *see* environmental degradation; subsidies
  and fire, nature-positive contribution of 7–8, 123–4, 131
  land, private 100, 101, 121

hailstorm: Kachchh, Gujarat, India 39–40, 44, 45–6
'high-reliability' individuals and practices 3–4, 130
Hinduism 42, 44
humanitarian aid 87–8, 89, 122, 129–30
hybrid rangeland governance 11, 15, 51–62, 120

income
  diversification 15, 67, 75, 85–6, 89–90
  remittances 111
  technology treadmill trap 69, 72–4
  see also milk production; wealth differences
insurance *see* Borana, southern Ethiopia; gender; livestock insurance; local responses and insurance, combining; younger generation
investments 128–9
Isiolo, northern Kenya 12–13, 23, 24, 79–84
  actively embracing uncertainty 87–8

adaptive technology 89
identities, relationships, and collaboration 88–9
Isiolo Kenya map *82*
networking, trust, and diversification 85–90
pastoral practices and external interventions 86–90
redistribution and resource pooling 84–5
role of pastoralist moral economies 80–1
wildlife attacks on livestock 29–30, 84
Islam and role of mosque, northern Kenya 84

Kachchh, Gujarat, India 9–10, 23, 26, 39–41
embracing uncertainty 44–7
limits to adaptations 47–9
Kachchh map *117*
mobility 30–1, 42, 43, 44
Rabari: context and practices 42–4

lake expansion 28–9, 53–4, 55–6, 59–60
land
access, negotiated 11, 67, 76, 113–4
degradation 5, 7, 12, 56, 121
grabbing 1, 6–7, 12, 16, 122, 126
reform 4, 67, 71
tenure 15, 57–8, 79, 109
see also hybrid rangeland governance; grassland; grazing
leadership role (*mukhis*) 46
'liquid modernity' 48, 126
livelihood diversification 15, 67, 75, 85–6, 89–90
livestock insurance
Kenya 89
see also Borana, southern Ethiopia
livestock management 15
pastoral vs industrial 6–7
see also specific case studies
livestock-raiding: northern Kenya 84, 85

local knowledge and experience 46, 52–3
see also visual methods
local responses and insurance, combining 98–102
locust plague: southern Ethiopia 27–8

markets 127–8
Middle East and North Africa: pastoral policies 121
migration *see* mobility/migration
milk production
and cheese making, *see* Sardinia, Italy
price fluctuations 33, 68
wealth differences 25
mineral water factory 57–8
mobility/migration 4, 11, 14–6, 8, 65, 89, 95, 115–6, 125–9, 131, 133
and 'absence', southern Tunisia 32, 107, 109–12
Gujarat, India 30–2, 39, 41–48
Middle East and Africa 120–2
restriction during Covid-19 pandemic 23–4, 28, 89–90
southern Ethiopia 100–1
see also agriculture; gender
monasteries, role of 57–9
moral economies, role of 2, 13, 79–91, 104, 120, 130–1
mosques, role of 84, 87, 89

narrative
anti-livestock, colonial 6, 77, 129, 132
development 3, 68–73
policy 131–2
see also pastoral policies
nature-positive contribution to climate and biodiversity crises 6–8
networks 74, 75, 77, 88–9, 110–11
gendered 85–6, 89–90
religious *see* monasteries; mosques
social media/Facebook 23–4, 32
transnational 32
newspapers 34–5

non-equilibrium environments 5
Normalized Difference Vegetation Index (NDVI) 94

pastoral policies
  agenda for action 132–3
  narratives and change 122–30, 131
  regional overview 120–2
pastoralism and rangelands 1–2
  challenges for development 8–9
  characteristics of pastoral systems 15–16
  definition and importance of 4–6
  nature-positive contribution to climate and biodiversity crises 6–8
  reframing development 2–4
  see also hybrid rangeland governance
PASTRES programme 21–2
photo elicitation 24–6
photographs
  documentary video and 26–7
  rephotography 24
  see also visual methods
photovoice 22–3
  feedback sessions 29, 52, 55
political economy 16
  Gujarat, India 42–4, 47–8
  southern Tunisia 111
private enclosures, expanding 100, *101*

Rabari *see* Kachchh, Gujarat, India
rainfall *see* droughts; hailstorm
reliability 3–5, 10–3, 46, 69, 80, 87–90, 119, 125, 130–3
religion *see* Buddhism; Hinduism; Islam
remittances 111
rephotography 24
resilience/resilient 9, 13, 60, 80–3, 86–91, 115–7, 119, 122, 124, 130–2
resource competition, consequences of 9
resource pooling 85, 112–15
rewilding 7

Sardinia, Italy 11–12, 25, 26, 33, 65–7
  common uncertainties 66–7
  conceptions and management of uncertainty 33
  development narrative: efficiency and productivity 68
  flexibility of small-scale and direct sales 74–7
  living with uncertainty 68–9
  Sardinia map *70*
  semi-intensive and sedentary livestock production 69–74
satellite technology 94
Seeing Pastoralism exhibition, Sardinia 35–6
semi-intensive and sedentary livestock production 69–74
sheep *see* Douiret, southern Tunisia; Kachchh, Gujarat, India; Sardinia, Italy
sheep-shearing 25, 26, 32
small-scale pastoralists and direct sales 74–7
social media/Facebook 23–4, 32
social networks *see* networks
social protection 4, 13, 79–81, 86–7, 94, 122, 130, 132
social and political relations 15–16
South Asia: pastoral policies 120–1
state governance and hybrid arrangements 56–60
sub-Saharan Africa: pastoral policies 121–2
subsidies
  EU 67, 121
  grazing ban zone and 59, 60
  Middle East and North Africa 121

Tataouine *see* Douiret, southern Tunisia
technology
  adaptive 89
  and labour resources 85
  milking machines 68, 69
  satellite 94
  smartphones 48
  social media/Facebook 23–4, 32
  treadmill trap 69, 72–4
temporality 48–9

temporal and spatial flexibility 4, 42, 44–7, 48
tourism/tourist 6, 11, 56, 59, 66, 74, 78, 122, 128
township relocation 59, 60
transnational networks 32
tree-planting campaigns 7

uncertainty
   definitions and conceptions of 1–3, 12, 21–2, 33–4, 37, 53–6, 61
   see also Amdo Tibet, China; Douiret, southern Tunisia; Isiolo, northern Kenya; Kachchh, Gujarat, India; Sardinia, Italy; visual methods
United Nations International Year for Rangelands and Pastoralism (2026) 133

variability 7–8, 11–3, 15, 22, 41, 69, 119, 131–3
   high 3–5, 9, 16

managing 45–6, 66, 116
veterinary/medical services 71–2, 77, 112
visual methods 21–2
   afterlives and circulation of material 34–6
   conversations around uncertainty 27–33
   research design and methodological reflections 22–7
   summary and conclusion 36–7

water resources 13, 24
water and soil management: southern Tunisia 108–9, *110*
wealth differences 8–9, 25
   insured and uninsured pastoralists 96–102
wildlife attacks on livestock 29–30, 84

younger generation 48, 71, 73, 74, 85, 111, 112
   insurance 96, 98